Test Methods to Determine Hazards of Sparingly Soluble Metal Compounds in Soils

Test Methods to Determine Hazards of Sparingly Soluble Metal Compounds in Soils

Edited by

Anne Fairbrother
Parametrix, Inc.
Corvallis, Oregon, USA

Peter W. Glazebrook
Rio Tinto Ltd.
Bundoora, Melbourne, Australia

José V. Tarazona
Spanish National Institute for Agricultural and Food Research
and Technology (INIA)
Madrid, Spain

Nico M. van Straalen
Vrije Universiteit
Department of Ecology and Ecotoxicology
Amsterdam, Netherlands

Current Coordinating Editor of SETAC Books
Andrew Green
International Lead Zinc Research Organization
Department of Environment and Health
Research Triangle Park, NC, USA

Publication sponsored by the Society of Environmental Toxicology and Chemistry
(SETAC) and SETAC Foundation for Environmental Education

Indexing by IRIS

Library of Congress Cataloging-in-Publication Data

Test methods to determine hazards of sparingly soluble metal compounds in soils / edited by Anne Fairbrother...[et al.].
    p. cm.
    Includes bibliographical references and index.
    ISBN 1-880611-42-2 (alk. paper)
        1. Metals--Environmental aspects--Testing. 2. Soil pollution--Environmental aspects--Testing. I. Fairbrother, Anne.

QH545.M45 T47 2002
628.5'5--dc21

20002022676

Information in this book was obtained from individual experts and highly regarded sources. It is the publisher's intent to print accurate and reliable information, and numerous references are cited; however, the authors, editors, and publisher cannot be responsible for the validity of all information presented here or for the consequences of its use. Information contained herein does not necessarily reflect the policy or views of the Society of Environmental Toxicology and Chemistry (SETAC®).

© 2002 Society of Environmental Toxicology and Chemistry (SETAC)
This publication was printed on recycled paper using soy ink.
SETAC Press is an imprint of the Society of Environmental Toxicology and Chemistry.
No claim is made to original U.S. Government works.
International Standard Book Number 1-880611-42-2
Printed in the United States of America
06 05 04 03 02   10 9 8 7 6 5 4 3 2 1

∞   The paper used in this publication meets the minimum requirements of the American National Standard for Information Sciences — Permanence of Paper for Printed Library Materials, ANSI Z39.48-1984.

Reference Listing: Fairbrother A, Glazebrook PW, Tarazona JV, van Straalen NM, editors. 2002. Test Methods to Determine Hazards of Sparingly Soluble Metal Compounds in Soils. Pensacola, FL, USA: Society of Environmental Toxicology and Chemistry (SETAC). p 128.

SETAC Publications

SETAC publications are intended to provide in-depth reviews and critical apprais-
als on scientific subjects relevant to understanding the impacts of chemicals and
technology on the environment. The books explore topics reviewed and recommended by the Publications Advisory Council and approved by the SETAC Board of
Directors for their importance, timeliness, and contribution to multi-disciplinary
approaches to solving environmental problems. The diversity and breadth of
subjects covered in the publications reflect the wide range of disciplines encom-
passed by environmental toxicology, environmental chemistry, and hazard and risk
assessment. These volumes attempt to present the reader with authoritative
coverage of the literature, including paradigms, methodologies, and controversies;
research needs; and new developments specific to the featured topics. The books
are generally peer reviewed for SETAC by acknowledged experts.

SETAC publications, which include books, Technical Issue Papers (TIPs), workshop
summaries, newsletter (*SETAC Globe*), and journal (*Environmental Toxicology and
Chemistry*), are useful to environmental scientists in research, research management, chemical manufacturing and regulation, risk assessment, and education, as
well as to students considering or preparing for careers in these areas. The publications provide information for keeping abreast of recent developments in familiar
subject areas and for rapid introduction to principles and approaches in new
subject areas.

SETAC would like to recognize the past SETAC Books editors:

       C.G. Ingersoll, Midwest Science Center

             U.S. Geological Survey, Columbia, MO, USA

       T.W. LaPoint, Institute of Applied Sciences

             University of North Texas, Denton, TX, USA

       B.T. Walton, U.S. Environmental Protection Agency

             Research Triangle Park, NC, USA

       C.H. Ward, Department of Environmental Sciences and Engineering

             Rice University, Houston, TX, USA

Contents

Chapter 1
Introduction 1
Anne Fairbrother, Guy Ethier, Mikael Pell, José V. Tarazona, Hugo Waeterschoot

Chapter 2
Soil Chemistry 5
Michael J. McLaughlin, Rebecca E. Hamon, David R. Parker, Gary M. Pierzynski, Erik Smolders, Iain Thornton, Gerd Welp

Chapter 3
Recommendations for Testing Toxicity to Microbes in Soil 17
Stephen P. McGrath, Ronald T. Checkai, Janeck J. Scott-Fordsmand, Peter W. Glazebrook, Grame I. Paton, Suzanne Visser

Chapter 4
Soil Toxicity Tests—Invertebrates 37
Hans Løkke, Colin R. Janssen, Roman P. Lanno, Jörg Römbke, Sten Rundgren, Nico M. van Straalen

Chapter 5
Terrestrial Plant Toxicity Tests 59
Frank van Assche, Jose L. Alonso, Lawrence A. Kapustka, Richard Petrie, Gladys L. Stephenson, Robert Tossell

Chapter 6
Summary and Conclusions 83
Anne Fairbrother, Guy Ethier, Mikael Pell, José V. Tarazona, Hugo Waeterschoot

List of Figures

List of Tables

Acknowledgments

The participants thank Dr. José V. Tarazona and his staff in Madrid, Spain, for help with local arrangements, the Society of Environmental Toxicology and Chemistry (SETAC) for support in organizing the meeting, and the remainder of the SETAC staff, particularly Greg Schiefer, Ria Vogelaerts, and Linda Longsworth, for assistance in meeting planning and report production.

We would also like to acknowledge our colleagues, Steven Siciliano and Bill Adams, who provided the peer review for this book. Lastly, we would like to acknowledge the work and support of the SETAC Office and staff in the production of this book.

Financial Contributors

International Copper Association (ICA)

International Council on Metals and the Environment (ICME)

International Lead Zinc Research Organization (ILZRO)

International Zinc Association

Nickel Industry Producers' Environmental Research Association (NiPERA)

Rio Rinto Ltd.

Union Miniere

U.S. Environmental Protection Agency (USEPA)

About the Editors

Anne Fairbrother, DVM, PhD, is Senior Wildlife Ecotoxicologist and Certified Wildlife Biologist with Parametrix, Inc. and holds adjunct professorship at the College of Veterinary Medicine of Oregon State University. She has degrees in Wildlife Management and Veterinary Medicine from University of California-Davis and a Masters and PhD in Wildlife Disease Ecology from University of Wisconsin-Madison.

Dr. Fairbrother has worked for the Environmental Research Laboratory (U.S. Environmental Protection Agency) and ecological planning and toxicology, inc. She was part of the management team and provided administrative support for programs in Wildlife Ecotoxicology, Superfund Site Assessment, and Biotechnology research and led studies in applied ecology relevant to terrestrial wildlife.

She served on the Advisory Panel on Ecotoxicity of the International Lead Zinc Research Organization, Nickel Producers' Environmental Research Association, and International Copper Association; the USEPA's Center of Excellence in Ecotoxicology Science Advisory Committee; and the National Research Council Committee on Animals as Sentinels of Environmental Health Hazards. She was President of the American Association of Wildlife Veterinarians and the Wildlife Disease Association and is on the Board of Directors of the Society of Environmental Toxicology and Chemistry (SETAC). She has served as Associate Editor for the *Journal of Wildlife Management*, and on the editorial boards of the *Journal of Wildlife Diseases, Ecotoxicology*, and *Environmental Toxicology and Chemistry*.

Peter W. Glazebrook, PhD, MSc, is Principal Advisor of Toxicology at Rio Tinto Technical Services, Melbourne, Australia. He holds a PhD and MSc in Environmental Science from the University of Queensland. His industry experience includes the toxicological characterization of wastes, the remediation of contaminated sites and associated health, and ecological risk assessments.

He has advised both industry and government on new regulatory developments, such as the Australian National Pollutant Inventory, through invitation to scientific advisory committees. For the past three years he also has served on the Expert Advisory Panels of the Australian Research Council for Environment/ Engineering and Biological/Medical. Current interests include the application of the life-cycle approach to the mining and metals industry.

José V. Tarazona, DVM, PhD, is the Director of the Department of the Environment at the Spanish National Institute for Agricultural and Food Research and Technology (INIA). His activity has combined basic research and application of scientific results in ecotoxicological regulations. He is actively involved in the environmental hazard identification and risk assessment of chemicals. He has been a member of the European Union Scientific Advisory Corpus since 1992 and currently is vice-chairperson of the Scientific Committee on Toxicology, Ecotoxicology and the Environment, where he has chaired the working group on Effects and Risk Assessment for Terrestrial Ecosystems.

He has authored more than 100 scientific papers, has been involved in the organization of several international congresses and workshops, and has been external expert-consultant for international organizations including Organization for Economic Cooperation (OECD), World Health Organization (WHO), and United Nations (UN) agencies.

Nico M. van Straalen has been a professor of Animal Ecology since 1992 at the Vrije Universiteit, Amsterdam, where he teaches evolutionary biology, ecology, and ecotoxicology. He was trained as a biologist with specializations in zoology, biomathematics, and biophysics. He completed a PhD study on comparative demography of springtails (defended in 1983) in which he showed how the life histories of these soil-dwelling insects diverge according to the average position of the species in the soil profile.

After receiving his PhD, van Straalen specialized in ecotoxicology and contributed to the promotion of soil invertebrates as test organisms for soil contamination. He published a framework for soil risk assessment based on species sensitivity distributions in 1989. His present interests involve the mechanism of adaptation to excess metal exposure. His recent publications show the structure of metallothionein in a model species of springtail and its role in genetically determined increased metal excretion in populations living in metal-contaminated soil.

Workshop Participants*

Jose L. Alonso
Spanish National Institute for
Agricultural and Food Research
and Technology (INIA)
Department of Plant Protection
Madrid, Spain

Ronald T. Checkai
U.S. Army
Edgewood Chemical and
Biological Center
Aberdeen Proving Ground, MD, USA

Guy Ethier
International Council on Metals and
the Environment
Ottawa, ON, Canada

Anne Fairbrother
ecological planning & toxicology, inc.
Corvallis, OR, USA

Peter W. Glazebrook
Rio Tinto Ltd.
Bundoora, Melbourne, Australia

Rebecca E. Hamon
University of Western Australia
Soil Science and Plant
Nutrition Group
Nedlands, Western Australia,
Australia

Colin R. Janssen
University of Ghent
Applied Ecology and
Environmental Biology
Ghent, Belgium

Larry A. Kapustka
ecological planning & toxicology, inc.
Corvallis, OR, USA

Roman P. Lanno
Oklahoma State University
Department of Zoology
Stillwater, OK, USA

Hans Løkke
National Environmental
Research Institute
Department of Terrestrial Ecology
Silkeborg, Denmark

Steven P. McGrath
Institute of Arable Crop Research-
Rothamsted
Herts, United Kingdom

Michael J. McLaughlin
Commonwealth Scientific and
Industrial Research Organization
Land and Water
Urrbrae, South Australia, Australia

David R. Parker
University of California
Soil and Water Sciences Section
Department of Environmental
Sciences
Riverside, CA, USA

Graeme I. Paton
Aberdeen University
Department of Plant and Soil Science
Aberdeen, Scotland

Mikael Pell
Swedish University of
Agricultural Science
Department of Microbiology
Uppsala, Sweden

Richard Petrie
U.S. Environmental
Protection Agency
Washington, DC, USA

Gary M. Pierzynski
Kansas State University
Department of Agronomy
Manhattan, KS, USA

Jörg Römbke
ECT Oekotoxikologie GmbH
Flörsheim, Germany

Sten Rundgren
Lund University
Department of Ecology
Lund, Sweden

Janeck J. Scott-Fordsmand
National Environmental
Research Institute
Department of Terrestrial Ecology
Silkeborg, Denmark

Rick Scroggins
Environment Canada
Environmental Technology Center
Gloucester, ON, Canada

Erik Smolders
Laboratory of Soil Fertility and Soil
Biology
Heverlee, Belgium

Gladys L. Stephenson
ESG International
Guelph, ON, Canada

José V. Tarazona
Spanish National Institute for Agri-
cultural and Food Research and
Technology (INIA)
Madrid, Spain

Iain Thornton
Imperial College of Science,
Technology and Medicine
Center for Environmental Technology
London, United Kingdom

Robert Tossell
Geo Syntec
Guelph, ON, Canada

Frank van Assche
International Zinc Association
Brussels, Belgium

Nico M. van Straalan
Vrije Universiteit
Department of Ecology and
Ecotoxicology
Amsterdam, Netherlands

Susan Visser
The University of Calgary
Department of Biological Sciences
Calgary, AB, Canada

Hugo Waeterschoot
Union Miniere
Brussels, Belgium

Gerd Welp
Institute für Bodenkunde
Der Rheinischen Friedrich-Wilhelms-
Universität Bonn
Nußallee, Germany

* Affiliations were current at the time of the Workshop

Abstract

This report describes the results of four days of discussion and debate at the Hazard Assessment of the Metals in Soils Workshop sponsored by the Society of Environmental Toxicology and Chemistry (SETAC) and administered by the SETAC Foundation for Environmental Education, held 19-23 June 1999 in San Lorenzo de El Escorial, Spain. Workshop participants were divided into 5 workgroups: chemistry, microbe toxicity tests, invertebrate toxicity tests, plant toxicity tests, and applications. Hazard assessment applications include classification for the purpose of labeling of materials in commerce, soil criteria derivation, ecological risk assessment, and site remediation. The chemistry of metals differs from that of organic compounds, necessitating different approaches for hazard assessment.

Equilibration of metal-containing substances in soils during test setup is a major difference. A transformation protocol was proposed to simulate weathering of metal compounds in the environment to ascertain whether the bioavailability of these naturally persistent compounds might change over time and to estimate salt effects. Consensus on soil types was not achieved. Use of artificial versus natural soil, as well as the number and types of soils required, still remains unresolved for hazard classification. There was general agreement, however, that ecological risk assessment requires use of natural soils representative of the area under consideration, as our ability to extrapolate toxicity data across soil types is limited. The plant and soil invertebrate workgroups proposed standard species and described detailed test guidelines. Several microbial function tests were recommended by the microbial workgroup, but additional information will be required to standardize the methods and put the results into an ecological context. Determination of the potential for metals to biomagnify in the food chain is complex, as organisms have evolved various mechanisms to use, exclude, or take up these naturally occurring substances. Workshop participants strongly endorsed the concept that measurement endpoints chosen for all tests be ecologically relevant for both acute and chronic effects.

Executive Summary

This report is the result of a SETAC technical workshop held in Spain, 19-23 June 1999, that critically examined current standard soil toxicity test methods. Modifications to standard guidelines were proposed to accommodate the particular biogeochemical properties of metals and their derivative substances, including a discussion of the types of soils for which the data will be applicable, appropriate test species and their source and acclimation conditions, methods for incorporation of material into soil, and awareness of micronutrient requirements and interactions. The use of toxicity data in hazard identification, risk assessment, and criteria setting also was discussed in order to delineate clearly the context within which test guidelines may be applied.

Toxicity testing of substances containing metals and metalloids is distinctly different than testing other substances. To varying degrees, there are naturally occurring amounts of all metals in soils. These background concentrations often are significant relative to quantities added. Total metal loads remain constant in soils, although significant leaching losses are possible with certain combinations of trace elements and soil chemical conditions. In addition, significant changes in chemical form can occur over time, profoundly influencing the bioavailability and hence the toxicity of the metal. This can lead to complications, for example when deciding how to best calculate the effective dose. A related problem is the essentiality of several important metals (e.g., copper, manganese, molybdenum, zinc), which dictates that a complete dose-response curve would be comprised of deficient, adequate, and toxic ranges. On a practical level, however, deficiencies are unlikely to be a common problem, as background concentrations will provide adequate mineral nutrition in most short-term toxicity tests.

Soil Chemistry

Ideally, the same substrates would be used for all organism groups by all laboratories performing hazard assessment tests. The closest approximation to this situation might be the use of an artificial soil substrate. However, some organisms cannot grow in this substrate, and it is difficult to extrapolate results to normal soils. Natural "standardized" soils are available and protocols for collecting such soils are in place. However, the soil chemistry workgroup proposed that, rather than attempting to identify a soil or soils to be used by all laboratories, a range of chemical, physical, and toxicological characteristics for the test soils would be specified and each laboratory could collect appropriate soils for their own use.

For cationic metals, test soils should have a pH of 5 to 5.5 with low organic matter (OM) content (minimum of 2%). The texture of the soil should be a sandy loam to loam and the cation exchange capacity also should be low with the clay mineralogy dominated by 2:1 clay minerals. Iron, manganese, and aluminium (hydr)oxides should not be present in large quantities. For anionic metals, soils would have similar properties, except pH would be 7.5 to 8. A slightly calcareous soil would be acceptable. Moisture content should be maintained at 40% water holding capacity. A period of 2 to 7 days will be allowed after the mixing process before the introduction of the test organisms or, in the case of microorganisms, before the testing process begins. This period will allow some degree of equilibrium to be reached between the soil and the test substance. It is recommended that soils be stored at a temperature of 20 to 25 °C during both the equilibration and testing periods.

The hazard test procedure is conducted 2 to 7 days after mixing the substance into the soil and 60 days later. One set of soils also will be leached and partly dried at the start of the 60-day incubation period. The prolonged equilibration time provides an indication of the eventual fate of the test substance in soil. The leaching procedure simulates a scenario where weathering and dissolution processes release the toxic ingredient from a sparingly soluble substance. The leaching and equilibration times are limited for practical reasons, and therefore the reevaluated toxicity is not predictive of potential effects after many years of soil contact. Research is required to determine the appropriate equilibration and leaching protocol and to more carefully define the test soil parameters.

The results of hazard tests conducted in concert with the proposed equilibration and leaching protocols are insufficient for a risk assessment or site evaluation, since the tests are performed for only few soil types and two equilibration times. The results may be useful, however, to prioritize substances for risk assessment and to classify substances according to relative hazard.

Soil Microbial Tests

Soil microorganisms (fungi, bacteria, actinomycetes, and protozoa) can amount to 90% of the living matter in soils and contribute to essential soil functions such as decomposition of organic residues, nutrient cycling, and soil aggregation. Microbes provide the substrate for grazing soil fauna and form the base of the soil food chain. Because soil microbes are in direct contact with the soil particles and soil pore water that contain contaminants, they are ideal organisms for assessing effects of pollutants in a soil system. Although many different soil microbial tests exist, they are less standardized than soil invertebrate and plant toxicity tests due, in part, to the fact that soil microbial ecotoxicology is a relatively new discipline. There are no methods for quantitatively extracting microbial communities from soil in a manner reflective of the different conditions in various soils. Nor is it possible at present, to add microbial cultures to soil in a sufficiently standardized and ecologically relevant manner. Hence, the soil microbial workgroup recommends tests with soils containing indigenous microbial communities.

The workgroup recommends the standardization and use of 3 soil microbial tests: soil respiration (equivalent to carbon dioxide liberation), nitrate production, and bioluminescence. The proposed Organization for Economic Cooperation and Development (OECD) Guideline 217 carbon transformation test is the protocol closest to standardization for soil microbes. It can be used to determine the amount of microbial activity that is present in a soil contaminated with metals, providing an index of the potential for the metallic substance to affect one of the functions of microbial populations. The nitrogen transformation test can be standardized for metals as well and measures the influence of pollutants on nitrate production, another important soil function. Several bioluminescence-based biosensors are available. These tests use soil bacteria that luminesce ("light up") in the presence of particular chemicals. The amount of luminescence can be used as an indicator of the degree of soil contamination.

The workgroup recommended the standardization and use of 2 microbial function tests: soil respiration and nitrogen liberalization. These will indicate changes in microbial biomass as well as in nitrogen cycling capacity. The workgroup also proposed including a biosensing technique that can act as a rapid and sensitive indicator of the bioavailability of metals and the consequent impact on the target organisms. This test will be carried out by using water extracted from the soil. The workgroup recognized that these tests are not yet standardized, but they are close to being developed and becoming commercially available.

The workgroup reviewed 24 microbial methods for their suitability for assessing effects of metals to soil microorganisms. The methods are commonly used as research tools to measure carbon-, nitrogen-, phosphorus-, and sulphur-driven processes, as well as to assess soil microbial diversity. From this review, the workgroup concluded that the 3 tests mentioned above are the most suitable for routine hazard assessment of the soil microbial community.

Invertebrates

The soil litter is a complex ecosystem with a diverse taxonomy and its own set of trophic relationships. Soil invertebrates play an important role in energy-cycling pathways in soil systems and are a food resource for vertebrate wildlife species. Exposure of soil invertebrates to soil-associated contaminants may be through dermal uptake from the water in soil and/or through ingestion of soil particles. To date, standardized tests are available only for earthworms, although draft guide-lines are available for several other taxa. Consequently, this workgroup provided more in-depth descriptions of test methods. The soil invertebrate workgroup recognizes that soil invertebrate tests need to be conducted in natural soils, as the organisms require both food and structure that are not available in artificial soils. They endorsed the concept of recommending appropriate standard parameters for soils, such that laboratories would select their own natural soils but would result in similar metal bioavailability.

Recommended test species are earthworms (*Eisenia andrei*), potworms (*Enchytraeus albidus*), springtails (*Folsomia candida*), and predatory mites (*Hypoaspis aculeifer*). Taken together, these four species represent a diverse taxonomy and various feeding groups. Draft protocols exist for all these species, with standard protocols available for earthworms. The workgroup suggests that all such protocols include a range-finding test and they endorse the use of development of concentration-response relationships, rather than simply attempting to determine a toxicity threshold. This approach does not simply test whether or not chemicals are toxic, but rather at what level they become toxic. The endpoint of all four tests is the number of juveniles produced by the end of the experimental period. This is representative of the potential change in reproductive fitness caused by the substance of interest, which is an ecologically meaningful endpoint. The workgroup provided design specifications for conduct of tests with each of the proposed organisms as well as identifying required quality control measures. Both short-term and long-term methods are described.

If consistent, precise responses of soil organisms during toxicity tests with metals is desired, it is very important to be sure that previous contact with elevated concentrations of metals has not occurred. Increased metal tolerance can occur if test organisms are exposed to elevated concentrations of metals in their natural soil or in pretest cultures. Thus, the origin of test organisms used during the toxicity test must be documented, and a reference chemical test should be included at least biannually to verify that expected sensitivity still exists. Carbendazim was recommended as the reference chemical for earthworm and potworm tests and dimethoate was recommended for springtails and predatory mites.

Soil invertebrates may be used in combination with microorganisms for multiple species, community tests. Such tests presently are being developed and standardized. Field methods also are under development, particularly for assessing the functions of soil invertebrates (e.g., decomposition). However, such tests fall outside the scope of this discussion on hazard identification.

Concentrations of metals inside the bodies of invertebrates also may be considered indicators of effects, although for essential metals, measurement of total internal concentrations in invertebrates is of limited value because this concentration is usually regulated within a narrow range over a broad range of environmental concentrations. For nonessential elements, internal concentrations can be related to a critical internal threshold such as the lethal body concentration (known as the critical body residue concept). Currently, this type of information is available for only a few invertebrate species and is more applicable to site-specific risk assessments than it is to hazard classification. Nevertheless, potential transfer of metals to invertebrate-feeding higher animals (e.g., moles, shrews, thrushes) may become an important issue, and critical body residues have been suggested as an alternative approach to setting criteria for metals in soil.

The workgroup participants recognized that it would be appropriate to classify hazards of metal compounds to soil organisms by using the four proposed species and suggested that there is an urgent need to determine the ecological requirements of the test species in the selected test soils, including the effect ranges of the reference compounds in these soils. Research also is required to establish physiological ranges for essential metals in soil organisms so that deficiency syndromes can be avoided. The workgroup also suggested several ways in which hazard classification data could be of relevance to risk assessment. However, highly variable site-specific soil properties need to be standardized through appropriate algorithms to the test conditions, so realistic extrapolations that account for differences in bioavailability can be made.

Plant Bioassays

Plant tests are the most developed and have been in standard practice longer than tests described by other workgroups. The plant workgroup focused mainly on the testing requirements for hazard assessment, but included discussions of specific approaches for site-specific and generic risk assessment. Both short-term and long-term studies are needed, and endpoints related to growth, biomass, and reproduction should all be considered. However, for short-term testing, reproductive endpoints generally are not included as exposure periods longer than 14 to 21 days that may be needed, which would require nutrient amendments or greater soil volume. Therefore, the workgroup recommends a 14-day exposure duration for the purposes of hazard classification and longer exposure period for risk assessment. They found no information indicating that reproductive endpoints measured in long-term studies are more sensitive to metals than are vegetative endpoints measured in short exposures.

Both the OECD and the American Society for Testing and Materials (ASTM) have published standardized test protocols for plant toxicity assessment. Though the standardized tests collectively list 31 agronomic species suitable for toxicity testing, the open literature contains reports on phytotoxic responses to chemicals for more than 1,500 species from nearly 150 families. Procedures also are described for testing either amendments to soil or site soils that have varying levels of contaminants. New OECD guidelines currently under development will contain much more detail on test conditions, introduction of test substance, and tier testing. Generally, the standards have allowed maximum flexibility so study designs can be tailored to specific scenarios applicable for regional or site-specific assessments. However, hazard classification requires tests that are sensitive, reproducible, and relatively easy to perform. Therefore, these types of toxicity tests require greater standardization in terms of procedures, conditions, and test species. The workgroup supports a test that will provide information on both short-term and longer-term effects, including vegetative vigor and at lease one reproductive endpoint.

One of the most important variables influencing the relative hazard assessment of substances is the test matrix itself, i.e., the soil in which the plants are grown. Because of the need to reduce interlaboratory variability, the workgroup recommends establishing a very narrow set of soil parameters and developing an artificial test matrix easily formulated from globally available "natural" constituents. This will enable laboratories in different regions to access and generate similar soils for testing. The artificial soil originally proposed by the OECD for use in toxicity tests with earthworms led to the suggestion of an artificial soil matrix consisting 75% silica sand, 20% kaolinite, and 5% (peat or O-horizon) OM. Although the workgroup recognizes the disadvantages of using an artificial soil instead of a natural soil as proposed by the other workgroups, it is their opinion that the advantages outweigh the limitations for hazard identification.

Test soil preparation should begin 5 to 7 days before plant seeds are placed into the soils. A batch of artificial soil sufficiently large enough to satisfy the replicate requirements for a given treatment is amended with the test substance. The test soils for each treatment are divided into the test pots and allowed to sit and "equilibrate" for another 2 to 7 days, after which seeds are planted. The recommended test species are alfalfa (*Medicago sativa*), barley (*Hordeum vulgare*), radish (*Raphanus sativus*), and northern wheatgrass (*Agropyron dasystachyum*) or perennial ryegrass (*Lolium perenne*). Measurement endpoints for both the range-finding and definitive seedling emergence plant tests include percent emergence, root and shoot length, root and shoot wet mass and dry mass, and total wet mass or dry mass. The range-finding test is only 5 to 8 days duration, and, at a minimum, seedling emergence and root and shoot length should be measured. For the definitive 14 to 21 day plant tests, it is recommended that all endpoints be evaluated.

The range-finding and definitive tests differ sufficiently to warrant entirely different statistical designs. The objective of the range-finding test is to find the concentrations that elicit a response, whereas the definitive test describes what the concentration-response relationship looks like. A balanced analysis of variance (ANOVA) experimental design is recommended for the range-finding test, using a minimum of 6 treatments. The definitive test should follow an unbalanced regression design with 10 to 13 treatments. Statistical evaluation for the different measurement endpoints can be conducted by using probit regression procedures for metrics such as seedling emergence or survival and nonlinear regression procedures for continuous measures such as growth. A procedure to express multiple measurement endpoint responses as a single phytotoxic score also may be a useful means of representing hazard assessment data.

The soil chemistry workgroup proposed that chemical analysis of total extractable metal fraction be done at the beginning and end of the test. However, the plant workgroup found it highly desirable to measure exchangeable or exchangeable plus soluble fractions in the soil to get a better representation of the actual exposure to the plants. In addition to the soil characterization, the workgroup recommended that metal concentrations in plant tissues be measured to assess the toxic potential to biota that consume plants.

For regional or site-specific risk assessments, tests should be performed by using representative soils from the region or site of interest to account for bioavailability differences. Site-specific evaluations also may include assessment of phytoremediation technology (i.e., using plants to extract contaminants from the soil), necessitating the evaluation of appropriate plant species. Because of the difficulty of selecting "ecologically relevant" species out of the thousands of terrestrial flowering plant species, the workgroup recommends selecting surrogates that are representative of dominant life-forms (grasses, woody plants, etc.). Species that are present in both the most highly contaminated portions of the site and the portion of areas of low (or no) contamination should be considered. Procedures are available in the ASTM guidelines to assist on the use of native species. Assessing soil phytotoxicity may involve both a screening-level assessment based on the highest concentration of the target soil constituent concentration and a response assessment that includes multiple concentration effect relationships. Concentration gradients can be generated through diluting the contaminated soil with soils from reference areas. The same endpoints discussed for hazard evaluation should be used in these tests as well. Meteorological factors such as temperature, relative humidity, and light intensity and duration have a strong influence on the rate of water use by plants that pull soluble contaminants out of the soil as well. These parameters should be representative of the conditions found at the site of concern. Standard ANOVA and regression analyses can be used to determine effect thresholds and responses to contaminants of interest. Characterization of soil, plant, and water samples must be conducted to facilitate the evaluation of site-specific study objectives similar to that proposed under the hazard assessment. If the testing is conducted to evaluate tolerance of plants to elevated metal concentrations and to determine metal removal by plants, both soluble and total metal analyses should be conducted.

The workgroup recognizes the similarities among approaches for hazard determination, risk assessment, and site evaluation or remediation. The primary difference among the approaches is the recommended use of artificial soil matrix for hazard determination and site-specific soils for assessment purposes. The workgroup recommended conducting a comparative study to evaluate the repeatability and feasibility of plant toxicity tests on the two standard soils proposed by the chemistry workgroup and the artificial soil matrix, including an examination of the factors that influence bioavailability of metals to various plant species.

Conclusions

Potential hazards of metals and other substances to the soil ecosystem are coming under increasing scrutiny. Additionally, the European Union (EU) is beginning the process of developing a hazard classification system for materials in commerce relative to their potential to cause adverse effects in the terrestrial ecosystem. Standard hazard identification and site assessment protocols are needed for soil organisms and processes in support of these efforts. Clearly, specific properties of metals require different approaches than those used for testing organic compounds. Mixing and equilibration of metals in soils during test set up and "transformation" of the test substance over time are major differences. The use of the transformation protocol proposed by the chemistry workgroup would be equivalent to evaluating environmental persistence of organic substances for the purposes of hazard identification.

Soil type also is a large consideration for testing metals and although this topic was discussed at length, consensus and closure on this issue were not achieved. The use of a standard, artificial soil matrix was endorsed by the plant workgroup, while the soil invertebrate and microbial workgroups recognized a need for use of natural soils. Not all soils could be represented in standardized tests so parameters were recommended that maximized metal bioavailability without causing stress-related effects to the test organisms. There was general agreement, however, that ecological risk assessment requires hazard information developed from natural soils representative of the area under consideration due to current limitations on the ability to extrapolate toxicity data across soil types. Standardized protocols were described for plant and soil invertebrates, but the soil microbial workgroup recognized the lack of accepted standard methods and the need for additional research on interpretation and application of test results. It was acknowledged by all participants that there remains a need to identify and suggest further tests for soil ecosystem function, as well as for toxicity to above ground organisms from food-chain exposure or direct soil ingestion. The workshop participants strongly endorsed the concept that measurement endpoints chosen for all tests should be ecologically relevant for both short-term and long-term effects.

It became obvious from the workshop discussions that development of standardized test methods for hazard assessment has reached different stages in soil chemistry, soil microbiology, soil invertebrates, and plants. However, since soil quality is an integrated function of all four of these properties (as well as soil physics, which was not discussed at this workshop), the development of soil toxicity tests for each of these disciplines must occur collaboratively. Moreover, since conduct of all the proposed tests during a hazard identification program will generate large data sets, effective methods for evaluation of their ecological relevance must be developed. Important information may be missed if each test is evaluated in isolation. This suggests the need for a coordinated research program to develop an integrated strategy for hazard assessment of metals and metal compounds in terrestrial ecosystems.

CHAPTER 1

Introduction

Anne Fairbrother, Guy Ethier, Mikael Pell, José V. Tarazona, Hugo Waeterschoot

This book is the result of a Society of Environmental Toxicology and Chemistry (SETAC) technical workshop, Test Methods to Determine Hazards of Sparingly Soluble Metal Compounds in Soils, held in Madrid, Spain, 19-23 June 1999, that critically examined the current standard soil toxicity test methods. Modifications to standard guidelines were proposed in order to accommodate the particular biogeochemical properties of metals and their derivative substances, including discussion of the types of soils for which the data will be applicable, appropriate test species and their source and acclimation conditions, methods for incorporation of material into soil, and awareness of micronutrient requirements and interactions. The use of toxicity data in hazard identification, risk assessment, and criteria setting also was discussed in order to delineate clearly the context within which test guidelines may be applied.

Metals have some special properties that make it particularly difficult to quantify their level of hazard to soil-dwelling organisms, such as plants and invertebrates, or to the soil microbial community. Metals are naturally occurring, environmentally persistent substances; therefore, organisms have developed various adaptive mechanisms for dealing with their presence in the soil. Some organisms actively exclude many of the metals from being taken up. Others concentrate metals and sequester them in places that do not cause effects to the organisms but may make them less palatable to animals that consume them. Some of the metals (e.g., copper, cobalt, nickel, zinc) are required micronutrients for plants or animals. For these substances, organisms have developed homeostatic mechanisms to keep their body concentrations in the optimal range. If soil concentrations are low, these metals are actively taken up; at high concentrations, uptake mechanisms may be blocked or the metals may be excreted more rapidly. Finally, some metals (e.g., lead) have no known biological functions and, if present in sufficiently large concentrations, are toxic to organisms.

Furthermore, specific physical and chemical properties of metals make them difficult to work with when following standardized toxicity test protocols. Incorporation of sparingly soluble metal compounds into the soil is difficult, and retention of the material in soil over time significantly affects uptake and toxicity. In addition, soil properties that affect the availability requirements for and interactive effects of certain metals that must be in the soil for proper nutrition also have the potential to confound test results. Nevertheless, information about the hazard of metals and

metal compounds to soil organisms is necessary for labeling and handling of
materials in commerce, for assessing risks to terrestrial systems from new or
existing sources of metals, and for cleanup of contaminated sites.

Workshop Purpose and Goals

Many efforts on the national and international scales are underway to develop soil
quality criteria or to conduct either site-specific or regionally-based risk assessments for metals and their derivative substances in terrestrial environments. A
major component of any of these efforts is an accurate measurement of hazard,
which has been defined as the inherent capacity of a substance to cause an adverse
response in a living organism (OECD 1995). Additionally, the European Union (EU)
is moving forward with development of a classification system to label materials in
commerce and transport with respect to their potential hazard to the terrestrial
environment. Although there are numerous studies in the literature on toxicity of
metal or metal products to plants, animals, and soil microbes, most of the data are
inappropriate for comparative hazard ranking or for use in criteria development or
risk assessment because they were generated using nonstandard tests. Thus, the soil
matrix differs, a wide range of mostly agronomic species was used and a variety of
endpoints was measured. This results in a dataset that is difficult to use for regulatory purposes, as specification of hazard thresholds becomes impossible.

To date, there are only a few standard international guidelines for soil studies
although many others have been proposed (Léon and van Gestel 1994). The
Organization for Economic Cooperation and Development (OECD) has published
2 standardized tests: 1) earthworm acute toxicity test in artificial soil (OECD
1984a; ASTM 1997) and 2) early seedling emergence test, also in artificial soil
(OECD 1984b). The American Society for Testing and Materials (ASTM) and the
U.S. Environmental Protection Agency (USEPA) have standards for whole-plant
toxicity tests in soils (ASTM 1994, 1999; USEPA 1996a, 1996b). Test methods for
other soil invertebrates (e.g., nematodes) and for soil microbial processes have
been proposed but have not received any formal review by the OECD, EU, or
ASTM. Furthermore, none of the accepted or proposed test protocols address
particular problems of testing metals and their derivatives or other sparingly
soluble substances. For example, soil type, including such factors as organic matter,
pH and redox potential, and length of time the material is in the soil, profoundly
affects bioavailability and, therefore, resultant toxicity measures.

Data from toxicity studies are likely to be used for comparative hazard ranking,
establishing cleanup goals, and risk assessment. Although all three of these
management processes use hazard information, their applications are sufficiently
diverse that it is possible that toxicity tests would be designed differently for each
(Smith and Hart 1994). Hazard identification and ranking should be based on the
intrinsic properties of the substance. The real ability or likelihood that a substance

would cause detrimental effects to an ecosystem is the objective of a risk assessment or criteria-setting process and must account for additional factors such as production rates, environmental emissions, degradation and transformation rates, exposure pathways, different environmental conditions, bioavailability, etc. (Van Tilborg and van Assche 1995). Therefore, toxicity tests used for comparative hazard identification may be optimized using standard soils, while hazard information developed for risk assessment or criteria setting may use varied test parameters depending upon soil type or may require some method for adjusting a single measured toxicity value by appropriate soil parameters (e.g., pH and organic carbon) (Scott-Fordsmand et al. 1996).

Questions Addressed

The questions addressed at this workshop become 3-fold. First, how should existing standardized soil toxicity tests be modified to accommodate the particular properties of metals and inorganic substances such as required micronutrients, incorporation into soils of sparingly soluble substances, species adaptations and acclimatization, etc.? Second, what additional standardized tests are required to characterize hazards to important biotic components of the soil ecosystem? Third, how should the hazard assessment information on metal substances be integrated with exposure information (i.e., environmental chemistry, bioavailability, interaction with other naturally occurring metals or metalloids, etc.) to provide an assessment of risk to terrestrial environments, to develop universal soil criteria, or to develop comparative hazard identification?

Participation and Workshop Format

The workshop brought together 31 experts from 11 countries for 4 days of discussion on methods for assessing hazard of metals and metal compounds to soil organisms. Participants included regulators, academics, industry representatives, and consultants who face issues of how to develop or apply terrestrial hazard assessment concepts in the regular course of their profession (see Participant list). The uses and limitations of this type of hazard information in classification, risk assessment, and contaminated-site assessment were discussed as well. The workshop consisted of 5 workgroups focusing on

1) test methods specific to soil invertebrates,

2) soil microbial communities,

3) plants,

4) chemical properties of metals in soils, and

5) regulatory context in which hazard information for metals might be used.

References

[ASTM] American Society for Testing and Materials. 1994. Standard practice for conducting early seedling growth tests. In: Annual book of standards. Conshohocken PA, USA: ASTM. E1598-94.

[ASTM] American Society for Testing and Materials. 1997. Standard guide for conducting laboratory soil toxicity or bioaccumulation tests with the Lumbricid Earthworm *Eisenia fetida*. In: Annual book of standards. Conshohocken PA, USA: ASTM. E1676-97.

[ASTM] American Society for Testing and Materials. 1999. Standard guide for conducting terrestrial plant toxicity tests. In: Annual book of standards. Conshohocken PA, USA: ASTM. E1963-98.

Léon CD, van Gestel CAM. 1994. Discussion paper on the selection of a set of laboratory ecotoxicity tests for the effects assessment of chemicals in terrestrial ecosystems. Paris, France: OECD.

[OECD] Organization for Economic Cooperation and Development. 1984a. OECD guideline for testing of chemicals: Earthworm acute toxicity test. Paris, France: OECD. Guideline nr 207.

[OECD] Organization for Economic Cooperation and Development. 1984b. OECD guidelines for testing of chemicals: Terrestrial plants, growth test. Paris, France: OECD. Guideline nr 208.

[OECD] Organization for Economic Cooperation and Development. 1995. Test methods for hazard and risk determination and inorganic metal compounds. Paris, France: OECD.

Scott-Fordsmand J, Pedersen MB, Jensen J. 1996. Setting a soil quality criterion. TEN 3:20-24.

Smith P, Hart ADM. 1994. Comparison of ecological hazard/risk assessment schemes. Summary document for OECD workshop; 24-27 May 1994; Worplesdon, UK. Central Science Laboratory Ministry of Agriculture, Food, and Fish.

[USEPA] U.S. Environmental Protection Agency. 1996a. Seed germination/root elongation toxicity test. www.epa.gov/docs/OPPTS_Harmonized/ 850_Ecological_Effects_Test_Guidelines/Drafts/850-4200.pdf. Washington DC, USA: USEPA. EPA-712-C-96-154. 6 p.

[USEPA] U.S. Environmental Protection Agency. 1996b. Terrestrial plant toxicity, Tier 1 (revegetative vigor). www.epa.gov/docs/OPPTS_Harmonized/ 850_Ecological_Effects_Test_Guidelines/Drafts/850.4150.txt.html. Washington DC, USA: USEPA. EPA-712-C-96-163. 6 p.

Van Tilborg WJM, van Assche F. 1995. Integrated criteria document: Zinc; industry addendum. Rozendaal, Netherlands: International Zinc Association. 50 p.

CHAPTER 2

Soil Chemistry

Michael J. McLaughlin, Rebecca E. Hamon, David R. Parker, Gary M. Pierzynski,
Erik Smolders, Iain Thornton, Gerd Welp

Scope

M etals and other naturally occurring elements do not breakdown
and therefore remain in soils indefinitely (McGrath 1987),
although significant leaching losses are possible (McBride 1998). In addition,
changes in chemical speciation can occur over time (Barrow 1998; Hamon et al.
1998), and these changes can profoundly influence the solubility and bioavailability, and hence the toxicity, of the metal (Welp and Brümmer 1997). There may be
an assortment of dissolved forms of metal in the soil solution that differ with
respect to their bioavailability (Lorenz et al. 1997; Smolders et al. 1998). Furthermore, most metals are predominantly associated with the soil solid-phase (Anderson and Christensen 1988; Welp and Brümmer 1999). With some soil fauna,
trace-element exposure involves ingestion of the soil solid-phase so that the
effective dose is a complex function of the solid-phase forms and the chemistry of
the gut. There is little to suggest that currently available soil-chemical extraction
schemes can predict bioavailability to a wide range of species accurately. Finally,
some of the metals are commonly found in more than one oxidation state in the
environment. Chemical characteristics can vary markedly among these different
forms so that the toxicity varies substantially (Welp 1999), therefore, results based
on one oxidative state or ionic species cannot be directly compared to another.
This chapter examines these soil chemistry issues in detail and proposes appropriate test soils and methods of incorporation of test substances that increase the
realism of standardized testing of metal toxicity to soil organisms.

Unique Challenges Posed by Toxicity Testing of Metals

Background and essentiality

Toxicity testing of substances containing metals and metalloids is distinctly
different than testing of other xenobiotic substances. To varying degrees, there are
indigenous levels of all metals in soils. These background levels often are significant
relative to quantities added, although the indigenous and exogenous pools may

differ in bioavailability (see Persistence and ionic form). This can lead to complications, for example, when deciding how to best calculate the effective dose.

A related problem is the essentiality of several important metals (e.g., Cu, Mn, Mo, Zn), which dictates that a complete dose-response curve would be comprised of deficient, adequate, and toxic ranges. On a practical level, however, this is unlikely to be a common problem, because background levels will provide adequate mineral nutrition in most short-term toxicity tests. For microbial, invertebrate, and plant tests, however, it is important that the organisms are not subjected to nutrient stress or nutrient excess (especially for metallic micronutrients). A more significant problem occurs during subsequent interpretation of toxicity threshold values, for example when soil criteria are developed using large safety factors or statistical extrapolation methods such that allowable threshold values fall below normal background levels and into the deficiency range.

Persistence and ionic form

Total metal loads remain constant in soils, although significant leaching losses are possible with certain combinations of trace elements and soil-chemical conditions. In addition, significant changes in chemical form can occur over time, profoundly influencing the bioavailability and hence the toxicity of the metal. This is manifested in several ways. Most importantly, there may be an assortment of dissolved forms in the soil solution (the aqueous speciation). With terrestrial higher plants, the majority of evidence is consistent with the notion that toxicity can be ascribed to the concentrations (or more formally, the activities) of free, uncomplexed metal ions in soil solution (Parker et al. 1995). A few exceptions have been noted however. For example, Cd uptake by Swiss chard (*Beta vulgaris* L.) confirmed that the free ion is more available than Cl-complexed Cd, but the latter was not altogether "unavailable," i.e., it contributed measurably to plant Cd uptake (Smolders and McLaughlin 1996a, 1996b). Similar studies of sulfate complexation have suggested that the $CdSO_4^0$ complex is taken up just as readily as the free Cd^{2+} ion (McLaughlin et al. 1998). Microbial toxicity tests indicated that the complexation of Cd by water-soluble organic substances lowered its toxicity whereas the monovalent $Cd(OH)^+$ species formed under weak acidic and alkaline conditions had a stronger toxic effect than the free Cd^{2+} ion (Welp and Brümmer 1997).

Most metals are dominantly found in the soil solid-phase and in a variety of different chemical forms (exchangeable, specifically adsorbed, fixed or precipitated in both simple and complex minerals, etc.). Thus, solution phase metals are generally a small proportion of the total bioavailable pool of metals in soil, and the assessment of the availability of solid phase forms is an important part of hazard and risk assessment processes. The ease with which these forms can replenish the soil solution ions lost to biological uptake or leaching (i.e., the metal-buffering characteristic) varies widely with both trace element and soil-chemical characteristics. For some organisms, e.g., plants, exposure to metals in soil occurs predominantly through the soil pore water. However, with some soil fauna, trace-element exposure involves ingestion of the soil solid-phase, so that the effective dose is a complex function of the interaction of the solid-phase forms with the chemistry of the gut. There is little evidence to suggest that current available chemical extraction schemes can accurately predict bioavailability, especially for a suite of different trace metals across all organisms. Both individual extractants (e.g., water, diethylenetriaminepentaacetic acid [DTPA], ethylenediamin-N,N,N',N'-tetraacetic acid [EDTA], nitric acid [NH_4NO_3]) and sequential extraction procedures have been widely used to operationally define the "bioavailable" trace-element pool in soils, but these have never been shown to be an adequate surrogate for a direct biological assay of toxicity or uptake (Dudka and Chlopecka 1990; Welp 1999).

Oxidation state

Many of the metals have more than one oxidation number, which poses some additional complications. First, chemical characteristics, and thus toxicity, can vary markedly among different oxidation states. Therefore, the oxidation number of the trace element in a given substance must be known. This is not necessarily a trivial problem, as some materials could conceivably contain mixed oxidation states. Second, some oxidation states may be unstable in the soil environment, so that distinct changes in bioavailability may occur during even a short-term toxicity assay (e.g., Cr(III)/Cr(VI)).

Substances and Materials Discussed

The constraints on what type of materials may be evaluated for environmental hazard using the protocols recommended by this workshop are unknown. Most of the hypothetical examples discussed were simple metal salts (soluble and insoluble) or mixed substances (e.g., hydroxypyromorphite, Cu-Cr-As wood preservatives, etc.), but there is no obvious reason to exclude a number of waste materials, industrial byproducts, or other types of materials from classification regarding hazard to soil biota. However, for many industrial substances, there may be issues of material homogeneity as well as natural background concentrations of the metal constituents. It should be noted that the primary constraint may be pragmatic; the material must be of sufficiently small particle size and homogeneity that it can be conveniently blended with the reference soils.

Elements discussed

The suggested protocols described herein were developed primarily for substances containing the common "problem" metals: Cd, Cu, Ni, Pb, and Zn. These are the elements with which there is the most expertise in both soil chemistry and toxicity research. In general, the recommended protocols also should be suitable for evaluating Ag, As, Co, Cr, Hg, Mo, Se, Sb, Sn, U, Tl, and V. The use of protocols developed for metals may be applicable to other inorganic substances for hazard identification, except under circumstances where substance toxicity is the result of radioactivity or volatility.

General Constraints

Caution should be exercised in the use of this protocol for certain substances due to the following constraints: 1) hazard classification can only be increased as a result of the proposed transformation procedure and 2) the initial acute toxicity test may overestimate the hazard. This is because the proposed time for equilibration of the test substance in the soil is not based on specific knowledge of when the initial, rapid reaction of the substance with the soil has been completed or, for elements in which different valence states may be important, on knowledge of when valence-state stability has been achieved.

This proposed protocol is not suitable for hazard identification of substances in anaerobic environments or for site-specific risk assessments of metal-contaminated soils. There is no method to assess transformation of metals in soils over the long time frames associated with their environmental persistence. However this protocol does provide a method for the prioritization of relative hazard levels among substances.

Substrate Selection

Soils are heterogeneous mixtures composed of a large variety of organic, organo-mineral, and mineral components, as well as soluble substances. Therefore, the quality and quantity of contaminant binding is highly variable. The relation between the mobile and immobile contaminant pools is a function of the physico-chemical characteristics of the substance and the composition of the soils (especially contents of organic matter [OM]; Fe-, Al-, and Mn-oxides; clay), the soil pH, and the redox conditions. The soil acts as a buffer system for both organic and inorganic substances. This buffering consists of a variety of complexation, sorption, and precipitation processes at the outer and inner surfaces of the soil solid phase including solid-state diffusion. For soil microorganisms and higher plants, this means that soils can immobilize and reduce the exposure potential of harmful substances (e.g., metals) since only the fraction present in the soil solution phase can be taken up by microbes or higher plants. Immobilization may not be as complete for soil invertebrates, however, as they ingest soil particles and may be exposed through routes other than the soil solution.

Risk and hazard assessments for metals in the terrestrial environment have to take the buffer capacity of the substrate into account. Therefore, the choice of appropriate substrates for toxicity testing is of major concern. One immediate consideration is whether a common substrate should be used for the 3 organism groups (microbes, invertebrates, and plants). While it is not essential that a common substrate be used, such an approach potentially would yield more information than separate substrates. The ideal situation would have the same substrate or substrates used for all organism groups by all laboratories performing the hazard assessment test. The closest approximation to this situation might be the use of an artificial soil substrate, such as commonly used in some current toxicity tests (e.g., OECD 1984). Advantages for using an artificial substrate include preexisting acceptance in ecotoxicological testing, ease of preparation, and relatively homogeneous substrate over time and across laboratories. The disadvantages are

1) the substrate is not suitable for soil microbe toxicity testing or for some soil invertebrates;

2) the variations in chemical properties of the artificial soil constituents;

3) the use of kaolinitic clay rather than more typical 2:1 clays found in temperate zones;

4) the soil mix is devoid of Al, Fe, and Mn oxides present in most natural soils, these minerals being crucial in controlling metal bioavailability in soil; and

5) the difficulty in extrapolating results to natural soils.

Advantages for using natural "standardized" soils are that such soils are available (although they may not be suitable for these purposes) and protocols for collecting such soils are in place (Kuhnt and Muntau 1992). Further, such soils also represent a more realistic exposure scenario. Natural soils would be suitable for all three organism groups, and transformations of substances would be more realistic in natural soils than in an artificial substrate. Significant disadvantages are the difficulty in providing a homogeneous soil or soils to interested users and the large number of soils that would be required to represent the range of toxicologically relevant soil properties.

The compromise position suggested by the soil chemistry workgroup is to use natural soils, but rather than attempting to identify a soil or soils to be used by all laboratories, to specify a range of chemical, physical, and toxicological characteristics for the test soils and allow each laboratory to collect appropriate soils for their own use. The key normalizing factor would be that the ECx value of a specified reference toxicant (a soluble metal salt) would have to fall within a specified range for each of the test organisms.

Proposed test soils

Two general classes of metals and metalloids suggest the need for two general soil types for hazard assessment testing. Cationic metals would have higher bioavailability to plants, microbes, and soil invertebrates under conditions of low pH; low OM content; low Fe, Mn, and Al (hydr)oxide content; and low cation exchange capacity or clay content. Metals or metalloids occurring as oxyanions would have higher bioavailability under conditions of high pH; low OM content; low Fe, Mn, and Al (hydr)oxide content; and low clay content. The basic chemical, physical, and toxicological characteristics for two soils are described below. These are considered to be the minimum number and type of soils that should be used in hazard assessment. In both cases, it is assumed that the surface horizon is collected and used.

General characteristics

The soils should not be deficient in elements essential for plant growth, nor should they have concentrations of essential elements that are in excess of those necessary for normal plant growth. High levels of available phosphorus, for example, could react with soil-applied Pb and form very insoluble Pb phosphate thus influencing metal toxicity. Soils containing naturally high concentrations of metals, such as those developed from ultramafic rocks (igneous rocks that are dominated by ferromagnesian minerals), should be avoided. This should be confirmed by total elemental analysis. Soils should not have had any recent application of biocides, and should contain at least 1% of the organic carbon as microbial biomass.

Benchmark toxicological characteristics should be established before individual laboratories select the soils that they would use. The soils selected for the EUROSOILS project (Kuhnt and Muntau 1992) could be a logical starting point because the sites and the soils are already characterized in detail. Nevertheless, other sites with appropriate soils can be chosen. Once a suitable area has been identified, large quantities could be excavated, homogenized, processed, and stored. Conversely, for microbiological measurements where sample processing can influence results, field-moist soil could be collected as needed.

Cationic metals

A test soil that would accentuate the bioavailability of the cationic metals should have a low pH but not so low that significant microbial processes (e.g., nitrification) are inhibited or that Al or Mn toxicity becomes a concern to plants or soil invertebrates. Calciole organisms (organisms adapted to alkaline soil conditions) and/or Al-Mn sensitive species should not be used. The suggested $pH(H_2O)$ range is 5 to 5.5 (equivalent to a $pH[0.01\ M\ CaCl_2]$ of 4.5 to 5). The OM content should be as low as possible while still supporting viable flora and fauna populations (minimum of 2%). The texture of the soil should be a sandy loam to loam. These textural classes will ensure that a rather low clay content and that the soil will maintain desirable porosity and other soil physical properties after mixing. The cation exchange capacity also should be low with the clay mineralogy dominated by 2:1 clay minerals. Fe, Mn, and Al (hydr)oxides should not be present in quantities typical for lateritic soils.

Anionic metals

The soil that would maximize anion bioavailability would have similar properties to those used for cationic metals, with the exception of soil pH. The recommended $pH(H_2O)$ would be 7.5 to 8 (equivalent to a $pH[CaCl_2]$ of 7 to 7.5). A slightly calcareous soil would be acceptable (maximum $CaCO_3$ content 2%).

Hazard Evaluation Protocol

The soils to be used for the tests will be disaggregated and thoroughly mixed to ensure homogeneity. The mixing process will not involve grinding to a fine powder but will aim to preserve a crumb structure to provide a suitable habitat for soil invertebrates such as springtails. The moisture content should be maintained at 40% of water holding capacity (WHC) to prevent segregation during mixing.

The test substance will be mixed carefully with the soil, and may need to be prepared to a specific grain size (i.e., as a powder) prior to mixing. Massive metals will need to be ground to a specified particle size range for mixing with the test soils. The workgroup suggests using a size range of 0.5 to 1 mm to allow homogeneity of mixing. When appropriate, the test substance may be mixed with an inert carrier such as colloidal silica to facilitate the mixing process. Hazard assessments of metals generally will be based on the metallic substance or salt (i.e., zinc oxide, cadmium chloride, lead phosphate, metallic zinc), which may be soluble or sparingly soluble, and not on the metal single ionic form (e.g., Zn^{2+}).

The dose-response function should be expressed on a unit substance basis, i.e., mg substance per kg soil. Confirmation of the dose will be predicated on the analysis of a reference analyte in each substance (e.g., the metal suspected to be the most hazardous in the substance) or another analyte. The choice of the reference analyte should be based on an easily measured element present in high concentrations in the substance and low concentrations in the substrate. Standardized sampling guidelines and analytical methods will be used and will include quality assurance-quality control (QA-QC) procedures to determine both sampling and analytical accuracy and precision.

A period of 2 to 7 days will be allowed after the mixing process before the introduction of the test organisms or, in the case of microorganisms, before the testing process begins. This period will allow some degree of equilibrium to be reached between the soil and the test substance. It is recommended that soils be stored at a temperature of 20 to 25 °C during both the equilibration and testing periods.

Transformation Tests

A transformation protocol is proposed for 3 reasons:

1) Sparingly soluble metal compounds may dissolve in soil over time to release potentially toxic species to the bioavailable pool. Without a transformation test, hazards could be underestimated; bioavailable pools of metals decline rapidly after addition to soil.

2) In soils, strongly sorbed metals are less hazardous than weakly sorbed metals for the pore water exposure pathway, so that a hazard assessment derived from an aqueous toxicity test without a soil transformation protocol is not appropriate for the terrestrial environment.

3) Metals are often added to soils as compounds, where the accompanying ion may affect the hazard classification. For example, sulfate or chloride salts can markedly increase metal toxicity. Organic anions (ligands) may reduce metal toxicity. A transformation test is needed to ensure hazards from metals are not overestimated or underestimated due to accompanying ion effects.

Suggested methodology

The methodology suggested is only conceptual at this stage and requires experimental validation and adjustment. The steps suggested are not based on experimental results but on the known concepts and knowledge regarding potential reactions of metals and sparingly soluble metal compounds in soils.

The hazard test procedure is comprised of 3 parallel toxicity evaluations (Figure 2-1):

1) a test that is performed 2 to 7 days after mixing the substance into the soil,

2) a test that is performed 60 days after the 2 to 7 days initial incubation of the substance into the soil, and

3) a test where the soil is leached at 2 to 7 days after mixing the substance into the soil. The soil subsequently is partly dried and incubated for a similar total transformation time as in Test 2.

The choice of substrates, the mixing procedure, and the toxicity evaluations are performed similarly in all three tests. In this section, we elaborate only on the choice of aging time and on the leaching procedure.

A 2 to 7 day equilibration of the substance in soil prior to testing may result in an underestimate of its toxicity because of transformation reactions. Substance dissolution in soil may be a slow process, and toxicity may become apparent only after some longer equilibration time. Therefore, we suggest allowing the substance to transform and to reevaluate its toxicity after a defined equilibration time. If toxicity increases with time, then the substance will be considered as more hazardous. If, however, the toxicity decreases with time, no change will be made to the

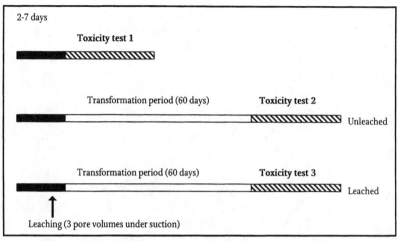

Figure 2-1 Proposed test procedure including equilibration and transformation periods

hazard classification of the substance since this classification is primarily intended for acute events (it is not really appropriate for longer-term risk assessment).

The prolonged equilibration time provides an indicative measure only of the eventual fate of the test substance in soil. The transformation time is limited for practical reasons, and therefore the reevaluated toxicity is only valid for relatively short time frames (e.g., is not accurately predictive of potential effects after many years of soil contact). The results of this proposed transformation test are not sufficient to make a risk assessment of the substance since the test is performed for only few soil types and two equilibration times. The proposed test is designed for hazard assessment only. The results may be useful, however, to prioritize substances for subsequent risk assessment (see section Links to risk assessment).

The transformation test is accompanied by a leaching procedure that simulates a scenario where weathering and dissolution processes release the toxic ingredient from a sparingly soluble substance. An unleached control is included also. If the leaching and transformation procedures change the toxicity compared to the initial test, then the results indicate that a transformation test is essential for a proper hazard assessment of the substance.

Figure 2-2 illustrates possible toxicity responses of soils tested immediately after substance addition (2 days) and soils with prolonged equilibration prior to testing (60 days) with and without leaching. In the illustration, toxicity decreases upon transformation, which, according to our definition, would not change the hazard classification of the substance (a precautionary approach for acute pollution events). The reverse also may occur, in which case the substance would be reclassified into a more hazardous category (Figure 2-3).

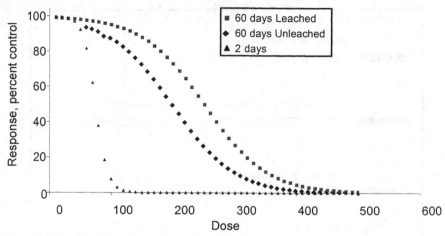

Figure 2-2 Possible effects of transformation and leaching on soil toxicity (metal subject to strong soil aging, accompanying ion being soluble and having additive toxicity)

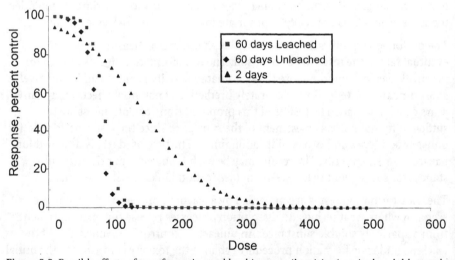

Figure 2-3 Possible effects of transformation and leaching on soil toxicity (sparingly soluble metal compound subject to significant dissolution in soil, accompanying ion effect small)

Equilibration time

The total equilibration time is 60 days after the initial 2 to 7 days incubation and prior to performing the toxicity test. The choice of the transformation time (60 days) is a trade-off between practical considerations and allowing a realistic amount of time for transformation reactions, with rapid reactions for metals typically having half-lives in the range of 1 to 100 days.

Leaching procedure

The leaching procedure should be performed in a soil column where leaching is made with 3 pore volumes of deionized water. The leaching is performed using suction (0.1 bar) at the bottom of the column. After the leaching, the soil is removed from the column and spread open to allow drying. Care should be taken that the soil moisture content does not drop below 40% of WHC. The soil then should be rewetted to the moisture content required by the test and should be further incubated to have a final residence time of the substance in soil of 62 to 67 days (similar to the unleached soil).

Research Needs

Suggested research to improve the proposed test protocol includes the transformation test period. The suggested length of the transformation period was based on approximate half-lives of transformation processes of common cationic metals in soils. Research is required to determine more accurately an appropriate duration in relation to the range of substances that might require evaluation. The nature of the organisms and biological endpoints used in the toxicity tests also may influence the duration of the transformation period. Research is needed to determine if two soils are sufficient to produce a valid hazard classification for the wide range of substances requiring evaluation.

The 2 soils selected were defined by a range of acceptable properties. Benchmark toxicity studies using a reference substance and the proposed test organisms are needed to evaluate the amount of variability introduced into the concentration response by the range of acceptable parameters. This may lead to a refinement and narrowing of the suggested ranges.

Links to risk assessment

Risk assessment of a substance involves both exposure and effects assessments in the environment for which the assessment of risk is made. Furthermore, the methodology to extrapolate from short-term laboratory toxicity tests to long-term environmental risk is tenuous. Risk assessment for metal substances in the environment requires assessment of critical exposure pathways for each receptor of concern (e.g., species of plants, animals, microbes) and appropriate estimates of background and bioavailability for each environment (e.g., ecosystem, mellalo-metallo-region). Information on the long-term fate of the substance is particularly important and is not provided by the hazard assessment methods proposed here.

The hazard classification protocol described above also should not be used for setting soil standards since the fate of the compound is tested in only 2 soil types and after only 2 equilibration periods.

References

Anderson PR, Christensen TH. 1988. Distribution coefficients of Cd, Co, Ni, and Zn in soils. *J Soil Sci* 39:1:15-22.

Barrow NJ. 1998. Effects of time and temperature on the sorption of cadmium, zinc, cobalt, and nickel by a soil. *Aust J Soil Res* 36:941-950.

Dudka S, Chlopecka A. 1990. Effect of solid-phase speciation on metal mobility and phytoavailability in sludge-amended soil. *Water Air Soil Pollut* 51:153-160.

Hamon RE, McLaughlin MJ, Naidu R, Correll R. 1998. Long-term changes in cadmium bioavailability in soil. *Environ Sci Technol* 32:3699-3703.

Kuhnt G, Muntau H, editors. 1992. Eurosoils: Identification, collection, treatment, characterization. Ispra, Italy: Joint Research Center, European Commission. Spec publication nr 1-94-60.

Lorenz SE, Hamon RE, Holm PE, Domingues HC, Sequiera EM, Christensen TH, McGrath SP. 1997. Cadmium and zinc in plants and soil solutions from contaminated soils. *Plant Soil* 189:21-31.

McBride MB. 1998. Soluble trace metals in alkaline stabilized sludge products. *J Environ Qual* 27:578-584.

McGrath SP. 1987. Long-term studies of metal transfers following application of sewage sludge. In: Coughtrey PJ, Martin MH, Unsworth MH, editors. Pollutant transport and fate in ecosystems. Oxford, UK: Blackwell Scientific Publications. p 301-317.

McLaughlin MJ, Andrew SJ, Smart MK, Smolders E. 1998. Effects of sulfate on cadmium uptake by Swiss chard: I. Effects on complexation and calcium competition in nutrient solutions. *Plant Soil* 202:211-216.

[OECD] Organization for Economic Cooperation and Development. 1984. Guideline for testing of chemicals: Earthworm acute toxicity test. Paris, France: OECD. Nr 207. 9 p.

Parker DR, Chaney RL, Norvell WA. 1995. Chemical equilibrium models: Applications to plant nutrition research. In: Loeppert RH, Schwab AP, Goldberg S, editors. Soil chemical equilibrium and reaction models. *Soil Sci Soc Am J* (Spec Publication) 42:163-200.

Smolders E, McLaughlin MJ. 1996a. Chloride increases cadmium uptake in Swiss chard in a resin-buffered nutrient solution. *Soil Sci Soc Am J* 60:1443-1447.

Smolders E, McLaughlin MJ. 1996b. Effect of Cl and Cd uptake by swiss chard in nutrient solution. *Plant Soil* 179:57-64.

Smolders E, Lambregts RM, McLaughlin MJ, Tiller KG. 1998. Effect of soil solution chloride on cadmium availability to swiss chard. *J Environ Qual* 27:426-431.

Welp G. 1999. Total and water soluble contents of nine metals in a loess soil causing dehydrogenase inhibition. *Biol Fertil Soils* 30:132-139.

Welp G, Brümmer GW. 1997. Microbial toxicity of Cd and Hg in different soils related to total and water soluble contents. *Ecotoxicol Environ Saf* 38:200-204.

Welp G, Brümmer GW. 1999. Adsorption and solubility of ten metals in soil samples of different composition. *J Plant Nutr Soil Sci* 162:155-161.

CHAPTER 3

Recommendations for Testing Toxicity to Microbes in Soil

Stephen P. McGrath, Ronald T. Checkai, Janeck J. Scott-Fordsmand, Peter W. Glazebrook, Graeme I. Paton, Suzanne Visser

Introduction

Soil is a living and dynamic system, consisting of a wide range of diversity of organisms such as plants, microbes, and invertebrates. Soil microbes are defined here as soil fungi, bacteria, actinomycetes, and protozoa, and such microorganisms can amount to 90% of the living matter in soils (Paul and Clarke 1989). Apart from this important contribution to the total soil biomass, they also greatly contribute to essential soil functions including decomposition of organic residues, nutrient cycling, and soil aggregation, which are vital to the functioning of the whole ecosystem (Torstensson et al. 1998). Microbes also provide the substrate for grazing soil fauna (Van Straalen and van Gestel 1994) and therefore are important constituents of food chains.

Because soil microbes are in direct contact with the soil particles and soil pore water that contain contaminants, they are ideal organisms for assessing the effects of pollutants in a soil system (Brookes 1993). Although many different soil microbial tests exist, they are less standardized than soil invertebrate and plant toxicity tests. For example, many of the international guidelines for microbial tests are recent while the earthworm and plant toxicity test guidelines are well established (OECD 1984a, 1984b; ISO 1993a, 1993b, 1995). The lack of standardization is due, in part, to the fact that soil microbial ecotoxicology is a relatively new discipline. Although the effects of pollutants on various components of the soil microbiota have been studied for a considerable length of time, and a diverse range of literature exists particularly in relation to amendment of soils with biosolids, the studies often did not focus on standard methods for hazard assessment of metal-containing substances.

Approach to microbial tests

There currently are no methods for quantitatively extracting microbial communities from soil in a manner reflective of the different conditions in various soils, nor is it possible, at present, to add microbial cultures to soil in a sufficiently standardized and ecologically relevant manner. Hence, to avoid problems of unknown

Test Methods to Determine Hazards of Sparingly Soluble Metal Compounds in Soils. Anne Fairbrother et al., editors.
©2002 Society of Environmental Toxicology and Chemistry (SETAC). ISBN 1-880611-42-2

selection of microbiota, the microbial workgroup decided to recommend tests with soils containing the indigenous microbial communities. This has the added benefit of mimicking the natural system, and including the natural diversity indigenous to a specific soil.

Test objective

The goal of the proposed tests is to assess hazards to microbial systems caused by metal-containing substances. This requires the identification of the most appropriate microbial parameters that can be used for detecting changes caused by such substances, in either their anionic or cationic forms. The workgroup decided that it was desirable to build on available draft standardized tests, or on those that could be standardized in the near future. However, new or nonstandard tests were reviewed as well, and recommendations were made as to their potential importance for hazard assessment to soil microbial community structure and function.

Review of Current Tests

A broad range of parameters has been used to assess effects of toxicants on microbial communities. As an overview of the tests available for evaluating microbial communities and processes in soils, a summary of the tests and an evaluation of their main features in relation to hazard assessment was made (Table 3-1).

Carbon-driven process assessment

Carbon mineralization has been well studied in relation to metal polluted soils (Scott-Fordsmand and Bruus 1995; Will and Suter 1995; Van Beelen and Doelman 1996). This can be assessed by either measuring either the evolution of CO_2 or the consumption of O_2. Barkay et al. (1986) claimed that the measurement of CO_2 was simpler, more rapid, and allowed the use of a much larger volume of soil, thus reducing problems associated with subsampling. Respiration can be measured by infrared gas analysis, gas chromatography, alkaline trap and back titration, or radio respirometry with a labeled pollutant. Problems can arise in the measurement of respiration, especially in carbonate rich soils that chemically evolve significant amounts of CO_2. However, Barkay et al. (1986) reviewed 26 studies and found only 3 measured O_2 consumption.

It also should be realized that carbon sequestration and CO_2 evolution will vary due to the nature of the stress, and therefore CO_2 evolution should not be used alone to assess changes in carbon utilization within a population. In fact, respiration may increase in a stressed soil (Brookes and McGrath 1984). However, it was agreed that this is an important process that is reproducible and relatively easy to measure. The substrate-induced respiration (SIR) method allows more detailed characterization of the soil response (Anderson and Domsch 1978), but the selection of substrate, as well as the concentration and method of addition of the

Table 3-1 Microbial parameters for soil ecotoxicity testing

Measurements	Reliability	Ease	Relevance	Sensitivity
Carbon-driven processes				
^{14}C plant residue decomposition	H	L	H	L
Basal respiration	H	H	H	L
SIR (single carbon source)	H	H	H	M
In situ C utilization (SIR with wide range of C sources)	H	L	H	M
Lag phase respiration	H	H	H	H
Biomass specific respiration (qCO_2)	H	H	H	M
Litter bag decomposition	H	H	H	L
Biolog® substrate utilization	L	L	L	L
Fumigation extraction (microbial biomass)	M	M	H	M
Nitrogen-, phosphorus-, sulphur-driven processes				
Nitrification	H	H	H	L
Denitrification	L	L	H	M
Mineralization of organic nitrogen	H	H	H	L
Nitrogen fixation (autotrophic)	L	L	H	H
Nitrogen fixation (heterotrophic)	L	L	L	H
Phosphorus, sulphur transformations	L	L	M	L
Diversity assessment				
Molecular measurements (wide range of techniques possible)	L	L	H	H
FAME[A]	L	L	H	H
PFLA[B]	L	L	H	H
Fluorescent probes	M	L	M	H
Conventional measurements				
Culturable organisms	M	M	M	L
Most probable numbers of cells	L	L	H	L
Tolerance (thymidine/leucine incorporation)	M	L	M	L
Xenobiotic degradation	M	L	M	Unknown
Enzyme- driven processes	L	H	L	M
Assays utilizing added organisms and aqueous extracts				
lux-based system	H	H	L	H
Microtox®	H	H	L	H
Protozoa	M	M	H	M

L = low, M = medium, H = high
[A]Fatty acid methyl ester profile
[B] Phospholipid-ester linked fatty acid analysis

test substance, must be considered carefully. The lag phase of CO_2 response following substrate addition in SIR can be determined easily and has shown to be a good indicator of metal stress (Sauvé et al. 1998).

There are benefits to combining respiration with other fundamental responses of the soil biomass. One benefit is that results can be manipulated to try and differentiate cell growth assimilation from actual respiration. It is beneficial to obtain the basal respiration, the lag time, and the specific respiration increment during the SIR measurement process (Stenström et al. 1998). Another benefit is that measurement of the specific rate of biomass respiration (qCO_2) has been shown to be a sensitive indicator of metal toxicity (Brookes 1993). The assumption of the SIR method is that organisms that grow during the incubation period are representative of the full spectrum of soil organisms (Anderson and Domsch 1978). However, there is no evidence to prove or disprove this assumption, which is problematic.

Two methods often have been used to assess carbon utilization by the microbial population. One is an extension of SIR to include a much broader range of carbon sources. Degens and Harris (1997) reported that this approach is sensitive in the assessment of land-use change. The technique however, has not been applied to changes associated with metals in soils. The extended test requires a great deal of analysis to be performed and is only likely to be routine when the measurement of CO_2 respired during the tests becomes automated.

A second alternative approach to detect changes in carbon utilization was described by Garland and Mills (1991). This method employs Biolog® plates originally developed for the identification of microbes in a medical context. The use of this technique has been broadened to reflect community level utilization of carbon sources. There now are a number of papers that evaluate the use of Biolog® plates to detect changes in metal contaminated soils (Garland and Mills 1991; Knight et al. 1995; Rutgers et al. 1998). The method is well automated but can be criticized from several perspectives. Primarily, Biolog® may not be an accurate indicator of soil microbial diversity since it is based on the range of carbon sources used by the culturable populations in a system. The method also requires great care in the normalization of the cell density by preassay plate counts to ensure consistency in the inoculum density applied to each of the wells on the plate. Finally, the data normally are analyzed by principal component analysis. Obviously, this does not lend itself to ECx values, so data analysis and presentation may need consideration prior to use in hazard identification.

Apart from the functional aspects of the soil carbon cycle, total microbial biomass carbon has been measured in metal-impacted soils (Brookes and McGrath 1984; Fleissbach et al. 1994). The biomass carbon component of the soil may change in 2 ways in the presence of metal pollutants: 1) reduction in numbers and species diversity within the total biomass may occur and 2) development of metal resistant microbial populations may take place. The fumigation method provides a reliable estimate of soil biomass concentration in metal contaminated and uncontaminated soils alike. This has been confirmed by the determination of adenosine triphosphate (ATP) concentrations and by direct microscopy techniques (Brookes et al. 1986).

Nitrogen-driven process assessment

Several studies have reported inhibition of nitrification in soils contaminated with metals (Giashuddin and Cornfield 1978, 1979; Chang and Broadbent 1982; Doelman 1986). However, measurement of fluxes involved in nitrogen cycling requires great care in the sampling and handling of the soil. Quantification of components of the nitrogen cycle requires the selective extraction of nitrogen from the soil matrix and the subsequent analysis thereafter. An appropriate extracting solution is one that has the following characteristics:

• allows quantitative extraction in a form that can be analyzed,
• inhibits biological and chemical reactions that may change the form of N,
• contains chemicals that will not interfere with the N assay, and
• remains stable for an adequate period of time.

The chemoautotrophic bacteria responsible for nitrification rely on sufficient O_2 and suitable moisture and temperature regime in addition to a supply of NH_4^+ and NO_2^-. Several heterotrophic organisms also have been implicated in nitrification (Haynes 1986; Killham 1986; Papen and von Berg 1998). Nitrification rates can be assessed by the measurement of ambient NO_3^- concentrations. Although nitrate ion specific electrodes offer rapid screening methods, the most commonly used quantitative technique involves nitrate determination using phenoldisulfonic acid and subsequent colorimetric analysis (Bremner 1965).

Phosphorus- and sulfur-driven process assessment

Although it would be ecologically relevant to measure impacts of metal compounds on the processes involved in transformation of sulphur and phosphorus, standardized methods for measuring sulfur and phosphorus processes are lacking. In the case of sulphur, the amounts mineralized would be relatively small and analysis for small changes in sulphur are chemically difficult (Freney 1979). Most of the phosphorus released after mineralization is largely immobilized in soil, making the measurement difficult and insensitive (Brookes et al. 1982). Assays of certain enzymes involved in the sulphur and phosphorus cycle are discussed below.

Enzymatic assessment

Enzymes catalyze numerous metabolic reactions in microbial cells and, accordingly, their inhibition could be the underlying cause of toxicity in a manner similar to that shown in many plants and animals. Activities of dehydrogenases, amylases, phosphatases, arylsulfatases, and cellulases have all been measured in metal polluted soils (Juma and Tabatabai 1977; Scott-Fordsmand and Pederson 1995; Will and Suter 1995). Enzymes in soil have been found to exist in at least 10 different forms depending upon their inhibition or stimulation (Burns 1982). The measurement of enzymes can provide a relatively inexpensive and rapid assessment of environmental samples.

Most studies have investigated the dehydrogenase enzymes and the phosphatase enzyme (Scott-Fordsmand and Pederson 1995). Specific dyes can be used in these assays to trace electron transport activity. The dyes act as artificial hydrogen acceptors and, upon reduction, change color allowing easy quantification with a spectrophotometer. However, in the case of Cu, chemical interference in the colorimetric assay can limit the usefulness of the dehydrogenase assay (Chander and Brookes 1991).

Although soil enzyme assays may offer high reproducibility, and in some cases relatively high sensitivity, the ecological relevance of the result is difficult to interpret. This is because the activity of these enzymes at any single point in time cannot be related directly to the fluxes of nutrients through soil. Many enzymes in soil are derived from plant roots as well as from microorganisms and are stabilized extracellularly; hence, enzymatic activity in soil at any one time is not a good measure of the microbial processes involved in nutrient cycling.

Measurement of ATP also has been applied in an attempt to assess metal pollution. ATP is a product of catabolic reactions and is destroyed rapidly after cell death, thus making it an ideal assay for differentiating living cells from dead cells. Typically, ATP is measured by the firefly luciferin-luciferase assay. Reasonably good correlations have been found between soil ATP concentrations and biomass carbon measured by other methods (Jenkinson et al. 1979; Oades and Jenkinson 1979; Brookes and McGrath 1984). Some researchers have estimated changes in the different adenosine pools (e.g., adenosine diphosphate [ADP], adenosine monophosphate [AMP]) and promoted this as a more ecologically suitable test system. Finally, the adenosine energy charge (AMP/ATP + ADP) has been used to show changes in energy levels in metal-impacted soil biomass (Vance et al. 1987). Nevertheless, the potential of using ATP content of actively growing cells to assess toxicity has seldom been systematically evaluated. The range of ATP kits that is available has made this assay convenient and relatively simple to carry out.

Biodiversity

Biodiversity measures emphasize community structure rather than activity or nutrient cycling processes. Both conventional and molecular techniques are available for evaluating soil microbial biodiversity.

Conventional measurements

Bacterial and fungal abundance have been estimated as colony forming units using plate count techniques. This method is particularly problematic for fungi because the plate count technique essentially measures spore numbers as actively growing hyphae have little chance to form a colony. It also is considered a doubtful method in the case of bacteria, fungi, and actinomycetes if rich media are used because only a fraction of the total organisms present will be able to grow on such media (Amann et al. 1995). To apply these techniques, it is essential to have a wide dilution series using a range of carbon sources, osmotic buffers, pH values, . temperatures, and various inhibitors. The selection of a specific extraction method (sonication, surfactants, etc.) also will favor certain components of the biomass. Furthermore, pollutants may change the component of the biomass that is viable but nonculturable and hence not detected by these conventional techniques.

Molecular techniques

Recently, more technically advanced methods have been applied to the measurement of diversity of microbial communities in soils. Soil microbiologists are aware that the unculturable organisms in soil may be important for essential soil processes. Molecular techniques that are presently being used to study changes in the nonculturable component of the soil include the fatty acid methyl ester (FAME) profile analysis, phospholipid-ester linked fatty acid (PLFA), and DNA extraction techniques.

The FAME method was initially developed for identifying single isolates but now is being applied at a community level. Using principal component analysis of the results, it has been possible to define similarities and differences between soil communities (Cavigelli et al. 1995). However, more importantly, when a species or group "signature" is detected, the analysis can become highly specific. The system is optimal when analyzing cellular components on "relatively" clean samples derived from pure isolates.

The PLFA method enables a higher degree of specificity than FAME (Bååth et al. 1998). Further developments of this method include the quantification of signature lipid biomarkers, which allows the discrimination between viable and "fossil" organisms (White et al. 1997). These techniques have potential because they are able to assess viability, community structure, and nutritional status. The possibility of linking alterations in microbial community structure to changes in carbon and nitrogen cycling processes may be greater with FAME and PLFA·techniques than with conventional methods. However organisms can change their fatty acid composition as a result of environmental factors, and this may limit the usefulness of these methods.

Extraction of DNA from soil has followed 2 approaches: direct and indirect methods. The direct approach yields on average an order of magnitude more DNA than indirect methods (Torsvik et al. 1995). However, the lack of quantitative recovery of DNA from soil has been recognized as a serious limitation (Steffan et al. 1988; Tebbe and Vahjen 1993; Cullen and Hirsch 1998). Although this is potentially a very powerful technique, the subsequent polymerase chain reaction on the extracted material also is open to criticism due to the formation of chimeric sequences that introduce variability at this stage of analysis (Liesack et al. 1991). Accordingly, these problems must be taken into account in the interpretation of the data. To resolve some of the shortcomings associated with DNA extraction and evaluation, some researchers have integrated several techniques to produce signature lipid markers (SLM)-DNA analysis. This approach is available commercially, but difficult to perform (Microbial Insights Inc., Tennessee, USA).

Like conventional techniques for assessing biodiversity, methods that consider the biodiversity of soil at the genetic level rely on the success of the extraction of the genetic material from soil. Various abiotic soil factors will determine the success and reproducibility of this process (Cullen and Hirsch 1998). Further, these techniques are highly specialized and are regarded at present as research tools rather than routine methods. This level of specialization makes them difficult to consider in the framework of this working group.

Assays using added organisms

Fungal bioassays have been developed to test effects of pollutants on specific fungi using such measurements as mycelial growth and productivity on solid media and in nutrient broth, spore germination, agar diffusion, respiration, or the measuring of potassium release from mycelium. These tests often do not have distinct endpoints and many are not conducted in intact soil. They therefore currently seem inappropriate for hazard screening procedures in the present context.

The effect of heavy metals on protozoa has been studied extensively in simple liquid systems and also in the soil environment (Foissner 1987; Forge et al. 1993). Protozoa may be sensitive indicators of soil pollution, and since many protozoa have short generation times, the effect of pollutants on populations of individual species or protozoan communities should manifest rapidly. The large size of most protozoa makes them easy to observe in soil or in pure culture. However, many of the assays used involve soil solution or soil extracts, making them inappropriate in the current study.

A number of marine organisms are naturally bioluminescent and can be both free-living and associated with higher marine organisms. Biochemically, bioluminescence is considered a branch of the electron transport system. The principle that a pollutant may inhibit metabolic activity and, as a consequence, decrease bioluminescence output, has been used for the detection of pollutants. Commercially available bioluminescent-based bacteria have been available for over a decade. The test-kit marketed by Azur (named Microtox®) has been internationally acknowledged as a sensitive, simple, inexpensive, reproducible, and rapid ecotoxicity assay in aquatic systems. Various reviews have considered the range of potential applications of the assay, including its extension to the soil ecosystem (Paton et al. 1997). Soil analyses are conducted either by extraction and testing of pore water or through a solid-phase analysis. Recently, the insertion of lux genes, which are responsible for bioluminescence, into a range of soil bacterial isolates has enabled the ease of measurement associated with bioluminescence to be combined with ecological relevance. This test is applicable across a broad pH range, and unlike the Microtox® assay, does not require the addition of a saline buffer that may alter the speciation of the metal in soil extract. However, most of the applications to date of lux-marked soil organisms involve exposure to isolated soil solution, thus limiting their relevance in the context of hazard evaluation.

Test Selection

The methods selected for studying soil microbial ecotoxicology focus primarily on microbial function rather than community structure for the following reasons:

- Measurements of microbial functions (processes) are relevant to all soils irrespective of their geographical location and pedo-ecoregion. Indeed, these measurements are directly related to the key nutrient cycling role of microbes in the soil and are critical to maintaining plant productivity.
- Microbial processes are relatively sensitive to metals in the soil.
- Measures of microbial function have well-defined endpoints, require little specialized equipment, and are cost-effective. They also are perceived as being highly reproducible.

Based on these criteria, the following techniques are listed for testing metal toxicity on microbial systems:

- carbon transformation test, including basal respiration, SIR, qCO_2, lag time, (Organization for Economic Cooperation and Development [OECD] Guideline 217 [2000]) and
- nitrogen transformation test (nitrogen mineralization of plant residue with low C:N ratio (OECD Guideline 216 [2000]; International Organization for Standardization [ISO] 14238 [1997]).

Carbon transformation test

The OECD Guideline 217 carbon transformation test (OECD 2000a) forms an excellent basis for determining dose or concentration-response of microbial carbon transformations in soil in the presence of metallic contaminants. This test generally is applicable to any metallic chemical except those that generate carbon dioxide abiotically. A brief summary of the method and recommendations as to how it would apply to hazard assessment of metal containing substances follows.

Soil selection and preparation

Selection of soils that maximize the bioavailability of metals to soil organisms is discussed in detail in Chapter 2. The OECD has established specific criteria for the test soil relative to the pH, texture, and amount of active biomass as a percentage of total organic carbon. Although the OECD recommends testing with only one soil, two soils are suggested for metal toxicity testing: 1) a loam or sandy loam, pH 5.3 to 5.6, containing 2 to 3% organic matter (OM), low in iron oxides and 2) a loam or silt loam, pH 7.5 to 8, 1 to 2% OM, and slightly calcareous. Microbial biomass carbon should be no less than 1% of total soil organic carbon as specified in OECD guidelines. It is important that the pH is not below 5 because microbial populations may not be supported in these conditions because of the presence of soluble aluminum.

Soils should be collected from areas that have a characterized land use as detailed in the OECD guidelines. Systems that have involved the widespread use of agro-chemicals are not acceptable. The soil should be sampled moist and should originate from the top 25 cm of the soil profile. After passing the soil through a 4 mm sieve, it can be stored at 4 °C for no more than 3 months. In countries where soil is frozen for more than 3 months each year, the test soil may be frozen and then stored for up to 6 months at −18 °C. Following storage, the microbial biomass should be measured to ensure it complies with the OECD guideline mentioned previously.

Prior to setting up the treatments, the stored soil should be equilibrated at ambient temperature (20 °C) for no less than 2 days. Moisture contents and water holding capacities (WHC) must be determined before initiating the test.

Testing procedure

A geometric series of at least 5 metal concentrations plus an unspiked negative control and reference positive control (e.g., zinc) should be tested. A benign salt control to assess the effect of trace nutrients associated with the anion may be required also, and, in that case, calcium is recommended. For water-soluble metallic substances, the carrier can be water applied as a spray or by pipette as stated in the OECD guideline (OECD 2000a). For sparingly-soluble compounds, the carrier will be the corresponding test soil, and, in this case, each test concentration will be formulated by the addition of spiked soil (no less than 5% w/w [weight for weight]). After adding the metal, each control and metal treatment is thoroughly homogenized and soil moisture adjusted to 40 to 60% of WHC as recommended in the OECD guideline. Moisture levels must be monitored and adjusted gravimetrically with deionized water throughout the term of the experiment.

Each of the metal-treated and control soils will be divided into 3 subsamples 48 hours following metal addition. One subsample will be leached immediately with deionized water to remove counter ions while the remaining 2 subsamples will remain unleached. Thus, for each test concentration and each control, 3 treatments will be tested: 1) short-term unleached, 2) long-term unleached, and 3) long-term leached. Following subsampling and leaching, all 3 subsamples within each metal and control treatment will be equilibrated for a further 5 days to allow soil microbial processes to stabilize.

Carbon transformation measurements will be conducted on all subsamples in the short-term unleached treatment to evaluate the immediate impact of metal contamination 7 days following metal addition. Basal respiration and SIR form the basis of the carbon transformation measurements but other dose or concentration-response measurements that are recommended include

- lag-phase (hours) in carbon dioxide response following glucose addition,
- specific growth rates,
- respiration of nongrowing and growing organisms, and
- metabolic quotient (qCO_2).

The protocol for measuring SIR is detailed in the OECD guideline. Long-term unleached and long-term leached treatments will be incubated for a further 60 days (total is 67 days) and then evaluated for carbon transformation potential as described above.

Data will be reported as the response expressed as a percentage of the negative control. Dose or concentration-response curves will be constructed for basal and glucose-induced respiration (GIR) rates after 7 and 67 days of exposure to the test metal. For each measurement, EC50 and EC10 values will be calculated, including 95% significance limits. The basic parameters for the curve for which these are calculated also should be provided.

Nitrogen transformation test

The OECD Guideline 216 (2000b) and ISO Standard 14238 (1997) for conducting the nitrogen transformation test are very similar and form a useful framework for testing dose or concentration response of the nitrogen transformation process to metallic contaminants. By incorporating modifications similar to those recommended for the carbon transformation test, these standardized tests can be applied to any metallic substance except those that may generate nitrate or ammonium abiotically.

Principle of the nitrogen transformation tests

The nitrogen transformation process involves the production of ammonium, nitrite, and nitrate from plant residues or OM as a result of microbial activities. In both the OECD and ISO test protocols, a nitrogenous substrate with a low C:N ratio (e.g., lucerne meal) is added to a control or treated soil, incubated for 28 days, and then analyzed for the production of mineralized N in a KCl extract of the soil. The endpoint for the OECD test protocol is nitrate production. Endpoints for the ISO test protocol are ammonium, nitrite, and nitrate production, and it is these endpoints that are recommended for determining microbial nitrogen transformations for the hazard assessment of metallic contaminants in soil.

Application of nitrogen transformation test to metal-contaminated soils

It is recommended that soil selection or preparation and the testing procedure follow the methods described for the carbon transformation test (see sections Soil selection and preparation and Testing procedure). Test treatments should include five metal concentrations plus an unspiked negative control and reference positive control as described previously. Immediately following the addition of the nitrogenous substrate and metal contaminants, the amounts of ammonium, nitrite, and nitrate must be determined using the KCl extraction method described in the ISO protocol. Extraction and measurement of ammonium, nitrite, and nitrate is repeated on the short-term unleached treatments 7 days following metal addition, rather than 28 days as described in the OECD and ISO guidelines. Differences between initial and 7 day nitrogen levels will elucidate the short-term impacts of metal contamination on the nitrogen transformation process. Ammonium, nitrite, and nitrate levels in the long-term leached and unleached treatments are measured again and compared to initial levels to determine the effects of extended metal exposure on the nitrogen mineralization process 67 days following metal addition. Data calculation, expression, and interpretation should follow that described for the carbon transformation test (see section Testing procedure).

Conclusions and Recommendations

After reviewing the techniques available for studying soil microbial ecotoxicology, it has been necessary to select 3 methods appropriate to assessing the hazard of metals in soils. The functional role of the soil microbial biomass was emphasized in the selection process because

- measurements of microbial functions (processes) are relevant to all soil irrespective of their geographical location and pedo-ecoregion, they are directly related to the key nutrient cycling role of microbes in the soil;
- measures of microbial function have well-defined endpoints, require little specialized equipment, and are cost-effective; they are also highly reproducible; and
- microbial processes are relatively sensitive to metals in the soil.

Soil respiration tests recently have been adopted by a range of international organizations and specific methodology has been proposed for the conduct of such assays. Both the carbon and nitrogen mineralization tests reflect the essential capacity of microbes to recycle plant material in soil. Problems associated with N-immobilization were considered, but the method adopted should resolve any of these issues by encompassing the following 4 salient points:

1) the low C:N ratio of the added material should reduce total immobilization;

2) the criteria for soil selection include the constraint that the biomass should account for at least 1% of the organic carbon in the soil;
3) there should be an absence of excess fertilizers and agrochemicals or other inhibitory chemicals in the soil prior to sampling; and
4) the environmental conditions in the soil should be optimal for the processes involved in the mineralization of organic N, the only limitation should be that imposed by the metal stress.

The carbon and mineralization tests should not be considered in isolation. The respiration assay should be compared with the microbial biomass and any changes in this parameter could be related to the rate of N-mineralization.

While the first two tests consider the response of processes in the soil, there is also a need for a more ecologically-based assessment. One such approach would be a measure of the microbial biodiversity of the soil system, but the literature available suggests that standardization of this could prove problematic. Biodiversity can be monitored only by using a molecular or genetic-based approach because of a large proportion of viable but nonculturable cells in the soil system. FAME, PLFA, and molecular techniques were reviewed but their use in metal toxicity assessment has not been standardized adequately. There also are problems both in the extraction stage and at the analysis stage, which have been reviewed earlier in this chapter. It is likely, however, that as these techniques become more routinely applied, they will begin to be more cost-effective. With subsequent standardization they could be considered as tools to compliment the current proposal.

Biosensors are a measure of the biological response of an organism to the bioavailability of a pollutant. As such, they are by definition direct indicators of a change in a biological impact of the solution phase. This is a unique characteristic that chemical tests cannot predict. So, in the future, they could be used to rapidly assess the progress of transformation of the added substances.

Limitations

There are a number of physiochemical attributes of the soil that must be characterized before and during the setup of the proposed assays. The biomass of the soil and its fraction of the percentage of OM must be maintained throughout the handling phases. Additionally, attributes such as the C:N ratio of the soil should be monitored routinely as an indicator of disruptive preparatory steps. If at any stage the biomass decreases during the pre-metal addition stage, the whole test will have to be restarted.

Regarding the actual tests, respiration poses some problems. First, any test involving measurement of CO_2 evolution can be affected by drying and rewetting of soil, giving a "flush" of CO_2—this should be avoided. Second, there needs to be some consideration of the method to assess CO_2 evolution. The measurement of CO_2 evolution is favored over other techniques (such as O_2 consumption) and has been investigated thoroughly. Alkali traps are recognized as overestimating biotic transformation (Rochette et al. 1997), automated respirometers are costly, and gas chromatography is highly time consuming. Problems associated with sampling and measurement can be resolved at a local scale with adequate quality controls in laboratories.

Regarding the N-mineralization test, there must be strict quality control in the source of the plant material. This material must have a known growth history such that there is no route for metals and/or other inhibitory substances to enter the bioassay. After extraction of the mineral N fractions from the soil, analysis of the fractions should be carried out quickly. There is a range of colorimetric and chromatographic techniques for the analysis of samples. The production of reproducible and reliable results will rely on adequate quality controls being adopted in individual laboratories.

The suggested tests do not consider molecular or genetic diversity because these techniques are highly specialized and may be difficult to interpret.

Research needs

As noted in the Review of current tests section in this chapter, many nonstandardized assays have been used to research effects of environmental stress on soil microbial communities. However, most of the tests require further research to understand the ecological relevance of the measurements or to standardize methodology.

It is emphasized that the goal of this publication was to explore ways in which the hazard of metals and metal-containing materials can be measured. Ecological risk-assessment was not the aim. However, this raises another important issue. More work is needed to show if the methods used for relatively short-term hazard assessment in the laboratory are related to what would happen in the natural environment. The main aspect that has been identified as important for microbial tests is the use of natural soils. This may add to the realism of the tests. It is essential to take this approach with indigenous soil microbial populations, communities, and diversity, as these cannot be recreated. This is in sharp contrast to previous plant and invertebrate studies where exogenous organisms were usually added, often to artificial soils.

However, in spite of the use of indigenous communities, the method of exposure used out of necessity in short-term laboratory tests is different than that normally experienced by microbial communities in the field. For example, in laboratory tests the entire dose of the substance being tested is typically added at one time. This type of contamination almost never occurs under field conditions, where contaminants usually slowly build up in soil. This gives the microbial populations time to adapt to the gradual increase in exposure. Outcomes may be different depending on the time available for adaptation, so the data from these tests may not be truly predictive of what will happen in nature. Depending on the form of the metal entering the ecosystem, its solubility may increase or decrease with time, interacting with the chemistry of the soil, again leading to different effects in short-term laboratory tests, compared with long-term impacts in the field. The tests are therefore most useful for hazard ranking of substances and can be criticized because they are not conducted for an adequate amount of time.

The selection of the 3 methods proposed here for hazard assessment of metallic substances was not an easy task; exclusion of a test should not be considered as reflecting upon the quality of the data from these methods. In addition, particular techniques that are worthy of further research include denitrification, inhibition of xenobiotic degradation (using ^{14}C), and certain molecular techniques, e.g., PLFA.

The 3 proposed tests also could benefit from additional research and development. For example, soil respiration following the addition of glucose to the soil matrix has been used in soil microbial ecotoxicity testing, but it would be valuable to verify this response against that measure following the addition of more ecologically relevant substrates. One option may be to use uniformly ^{14}C-labeled plant material and to monitor a mass balance of the breakdown, mineralization, and sequestration of the carbon. This response can then be compared to the suggested SIR approach to evaluate the efficacy of the test system.

The N-mineralization test would benefit from additional research as well. For instance, use of the stable isotope ^{15}N in the vegetative material of the N-mineralization experiment would provide a full understanding of the N flux. This is particularly important, because it will enable the "worst possible scenario" of immobilization of N to be assessed in systems that are particularly carbon rich.

The focus for additional research on biosensor assays must be towards the evaluation of the solid phase as an exposure surface for bacteria. If this approach is optimized, it will remove the requirement of the aqueous extraction step that currently prevents direct correlation of the biosensor assay with the other two proposed tests.

Interface with other groups

The soil is a living and dynamic habitat. Microorganisms do not survive in isolation of higher organisms, but have a range of antagonistic and symbiotic relationships with plants and invertebrates. A crucial aspect of the role of microorganisms relates to this interface, and it may be the case that this is where the sensitivity of microbes to environmental pollutants becomes most apparent (Killham and Firestone 1983; Koomen 1990; McGrath 1994; McGrath et al. 1995; Giller et al. 1998, 1999). Mesocosm-type experiments that consider the recommendations of all of the workgroups to integrate effects across various trophic levels may be the direction for future development of soil ecotoxicity tests.

References

Amann RI, Ludwig W, Schleifer KH. 1995. Phylogenetic identification and in situ detection of individual microbial-cells without cultivation. *Microbiol Rev* 59:143-169.

Anderson JPE, Domsch KH. 1978 A physiological method for the quantitative measurement of microbial biomass in soils. *Soil Biol Biochem* 10:215-221.

Bååth E, Frostegård Å, Díaz-Raviña M, Tunlid A. 1998. Microbial community-based measurement to estimate heavy metal effects in soil: The use of phospholipid fatty acid patterns and bacterial community tolerance. *Ambio* 27:58-61.

Barkay T, Shearer, DF, Olson BH. 1986. Toxicity testing in soil using micro-organisms. In: Dutka B J, Bitton G, editors. Toxicity testing using micro-organisms. Boca Raton FL, USA: CRC Press. p 134-154.

Bremner JM. 1965. Inorganic forms of nitrogen. In: Black C, Evans DD, White JL, Ensminge LE, Clark FE, editors. Methods of soil analysis. Part 2. Agronomy 9. Madison WI, USA: ASA. p 1179-1237.

Brookes PC. 1993. The potential of microbiological properties as indicators in soil pollution monitoring. In: Soil monitoring. Monte Veritá: Birkhäuser Verlag Basel. p 229-254.

Brookes PC, Heijnen CE, McGrath SP, Vance ED. 1986. Soil microbial estimates in soils contaminated with metals. *Soil Biol Biochem* 18:383-388.

Brookes PC, McGrath SP. 1984. Effect of metal toxicity on the size of the microbial biomass. *J Soil Sci* 35:341-346.

Brookes PC, Powlson, DS, Jenkinson DS. 1982. Measurement of microbial biomass phosphorus in soil. *Soil Biol Biochem* 14:319-329.

Burns RG. 1982. Enzyme activity in soils: Location and possible role in soil microbiology. *Soil Biol Biochem* 14:423-429.

Cavigelli MA, Robertson GP, Klug MJ. 1995. Fatty-acid methyl-ester FAME profiles as measures of soil microbial community structure. *Plant Soil* 170:99-113.

Chander K, Brookes PC. 1991. Is the dehydrogenase assay invalid as a method to estimate microbial activity in copper-contaminated soils? *Soil Biol Biochem* 23:909-915.

Chang FH, Broadbent FE. 1982. Influence of trace metals on some soil nitrogen transformations. *J Environ Qual* 11:1-4.

Cullen DW, Hirsch PR. 1998. Simple and rapid method for direct extraction of microbial DNA from soil for PCR. *Soil Biol Biochem* 30:983-993.

Degens BP, Harris JA. 1997. Development of a physiological approach to measuring the catabolic diversity of soil microbial communities. *Soil Biol Biochem* 29:1309-1320.

Doelman P. 1986. Resistance of soil microbial communities to heavy metals. In: Jensen J, Kjoller A, Sorensen LH. FEMS Symposium nr 33 Microbial Communities in Soil; 4-8 August 1985; Copenhagen, Denmark. London, UK: Elsevier Applied Science Publications Ltd. p 368-398.

Fliessbach A, Martens R, Reber HH. 1994. Soil microbial biomass and microbial activity in soils treated with heavy-metal contaminated sewage-sludge. *Soil Biol Biochem* 26:1201-1205.

Foissner W. 1987. Soil protozoa: Fundamental problems, ecological significance, adaptations in ciliates and testaceans, bioindicators, and guide to the literature. *Prog Protistol* 2:69-212.

Forg TA, Darbyshire JF, Berrow ML, Warren A. 1993. Protozoan assays of soil amended with sewage sludge and heavy metals using common soil ciliate *Colpoda steinii*. In: Eissackers HJP, Hamers J, editors. Integrated soil and sediment research: A basis for proper protection. Dordrecht, Netherlands: Kluwer Academic Publishers. p 315-316.

Freney JR. 1979. Sulfur transformations. In: Rhodes W, Fairbridge CW, editors. The encyclopedia of soil science. Stroudsbourg PA, USA: Dowden, Hutchinson, and Ross, Inc. p 536-544.

Garland JL, Mills AL. 1991. Classification and characterization of heterotrophic microbial communities on the basis of patterns of community-level sole-carbon-source utilization. *Appl Environ Microbiol* 57:2351-2359.

Giashuddin M, Cornfield AH. 1978. Incubation study on effects of adding varying levels of nickel as sulphate on nitrogen and carbon mineralization in soil. *Environ Pollut* 15:231-234.

Giashuddin M, Cornfield AH. 1979. Effect of adding nickel as oxide to soil on nitrogen and carbon mineralization at different pH values. *Environ Pollut* 19:67-70.

Giller KE, Witter E, McGrath SP. 1998. Toxicity of heavy metals to microorganisms and microbial processes in agricultural soils: A review. *Soil Biol Biochem* 30:1389-1414.

Giller KE, Witter E, McGrath SP. 1999. Assessing risks of heavy metal toxicity in agricultural soils: Do microbes matter? *Human Ecol Risk Assess* 5:683-689.

Haynes RJ. 1986. Nitrification. In: Mineral nitrogen in the plant-soil system. Orlando FL, USA: Academic Press, Inc. p 127-165.

[ISO] International Organization for Standardization. 1993a. Soil quality—Determination of the effects of pollutants on soil flora. Part 1: Method for the measurement of inhibition of root growth. Geneva, Switzerland: ISO. Standard 11269-1.

[ISO] International Organization for Standardization. 1993b. Soil quality—Effects of pollutants on earthworms *Eisenia fetida*. Part 1: Determination of acute toxicity using artificial soil substrate. Geneva, Switzerland: ISO. Standard 11268-1.

[ISO] International Organization for Standardization. 1995. Soil quality—Determination of the effects of pollutants on soil flora. Part 2: Effects of chemicals on the emergence and growth of higher plants. Geneva, Switzerland: ISO. Standard 11269-2.

[ISO] International Organization for Standardization. 1997. Soil quality—Biological methods. Determination of nitrogen mineralization and nitrification in soils and the influence of chemicals on those processes. Geneva, Switzerland: ISO. Standard 14238.

Jenkinson DS, Davidson DA, Powlson DS. 1979. Adenosine triphosphate and microbial biomass in soil. *Soil Biol Biochem* 11:521-527.

Juma NG, Tabatabai MA. 1977. Effects of trace elements on phosphatase activity in soils. *J Soil Sci Soc* 41:343-346.

Keeney DR, Nelson DW. 1982. Nitrogen-inorganic forms. In: Page AL, Miller RH, Keeney DR, editors. Methods of soil analysis. Part 2: Chemical and microbiological properties. Madison WI, USA: ASA. p 643-698.

Killham K. 1986. Heterotrophic nitrification. *Soc Gen Microbiol* (Spec Publication) 20:17-126.

Killham K, Firestone MK. 1983. Vesicular arbuscular mycorrhizal mediation of grass response to acidic and heavy metal depositions. *Plant Soil* 72:39-48.

Knight BK, McGrath SP. 1995. A method to buffer concentrations of free Zn and Cd ions using a cation exchange resin in bacterial toxicity studies. *Environ Toxicol Chem* 14:2033-2039.

Knight BP, Chaudri AM, McGrath SP, Giller KE. 1998. Determination of chemical availability of cadmium and zinc in soils using inert soil moisture samplers. *Environ Pollut* 99:293-298.

Koomen I, McGrath SP. 1990. Mycorrhizal infection of clover is delayed in soils contaminated with heavy metals from past sewage sludge applications. *Soil Biol Biochem* 22:871-873.

Liesack W, Weyland H, Stackenbrandt E. 1991. Potential risks of gene amplification by PCR as determined by [16]S rDNA analysis of a mixed culture of strict barophilic bacteria. *Microb Ecol* 21:191-198.

McGrath SP. 1994. Effects of heavy metals from sewage sludge on soil microbes in agricultural ecosystems. In: Ross SM, editor. Toxic metals in soil-plant systems. Chichester PA, USA: John Wiley and Sons, Inc. p 247-274.

McGrath SP, Chaudri AM, Giller KE. 1995. Long-term effects of metals in sewage-sludge on soils, microorganisms and plants. *J Indust Microbiol* 14:94-104.

Oades JM, Jenkinson DS. 1979. Adenosine triphosphate content of the soil microbial biomass. *Soil Biol Biochem* 11:201-204.

[OECD] Organization for Economic Cooperation and Development. 1984a. OECD guidelines for testing of chemicals: Earthworms acute toxicity test. Paris, France: OECD. Guideline nr 207.

[OECD] Organization for Economic Cooperation and Development. 1984b. OECD guidelines for testing of chemicals: Terrestrial plants, growth test. Paris, France: OECD. Guideline nr 208.

[OECD] Organization for Economic Cooperation and Development. 2000a. OECD guidelines for testing of chemicals: Soil microorganisms: Carbon transformation test. Paris, France: OECD. Guideline nr 217.

[OECD] Organization for Economic Cooperation and Development. 2000b. OECD guidelines for testing of chemicals: Soil microorganisms: Nitrogen transformation test. Paris, France: OECD. Guideline nr 216.

Papen H, von Berg R. 1998. A most probable number method MPN for the estimation of cell numbers of heterotrophic nitrifying bacteria in soil. *Plant Soil* 199:123-130.

Paton GI, Rattray EAS, Campbell CD, Meussen H, Cresser MS, Glover LA, Killham K. 1997. Use of genetically modified microbial biosensors for soil ecotoxicity testing. In: Pankhurst CS, Doube B, Gupta V, editors. Bioindicators of soil health. Wallingford, UK: CAB International. p 397-418.

Paul EA, Clark FE. 1989. Soil microbiology and biochemistry. San Diego CA, USA: Academic Press, Inc. 271 p.

Rochette P, Ellert B, Gregorich EG, Desjardins RL, Pattey E, Lessard R, Johnson BG. 1997. Description of a dynamic closed chamber for measuring soil respiration and its comparison with other techniques. *Can J Soil Sci* 77:195-203.

Rutgers M, van't Verlaat IM, Wind B, Posthuma L, Breure AM. 1998. Rapid method for assessing pollution-induced community tolerance in contaminated soil. *Environ Toxicol Chem* 17:2210-2213.

Sauvé S, Dumestre A, McBride M, Hendershots W. 1998. Derivation of soil quality criteria using predicted chemical speciation of Pb^{2+} and Cu^{2+}. *Environ Toxicol Chem* 17:1481-1489.

Scott-Fordsmand JJ, Bruus Pedersen M. 1995. Soil quality criteria for selected inorganic compounds. Silkeborg, Denmark: Danish Environmental Protecion Agency. Working Report 48.

Steffan RJ, Goksøyr J, Bej AK, Atlas RM. 1988. Recovery of DNA from soils and sediments. *Appl Environ Microbiol* 54:2908-2915.

Stenström J, Stenberg B, Johansson M. 1998. Kinetics of substrate-induced respiration: SIR theory. *Ambio* 27:35-39.

Tebbe CC, Vahjen W. 1993. Interference of humic acids and DNA extracted directly from soil in detection and transformation of recombinant DNA from bacteria and a yeast. *Appl Environ Microbiol* 59:2657-2665.

Torstensson L, Pell M, Steinberg B. 1998. Need of a strategy for evaluation of arable soil quality. *Ambio* 27:4-8.

Torsvik V, Daae FL, Goksoyr J. 1995. Extraction, purification, and analysis of DNA from soil bacteria. In: Trevors JT, van Elsas JD, editors. Nucleic acids in the environment. Berlin, Germany: Springer-Verlag. p 30-48.

Van Beelen P, Doelman P. 1996. Significance and application of microbial toxicity test in assessing ecotoxicological risks of contaminants in soil and sediment. Bilthoven, Netherlands: RIVM. Research for Man and Environment. Report nr 719102051.

Vance ED, Brookes PC, Jenkinson DS. 1987. An extraction method for measuring soil microbial biomass-C. *Soil Biol Biochem* 19:73-77.

Van Straalen NM, van Gestel CAM. 1994. Soil invertebrates and microorganisms. In: Calow P, editor. Handbook of ecotoxicology. Oxford, UK: Blackwell Scientific Publications. p 251-277.

White DC, Ringelberg DB, Macnaughton SJ, Alugupalli S, Schram D. 1997. Signature lipid biomarker analysis for quantitative assessment in situ of environmental microbial ecology. *ACS Symp Ser* 671:22-34.

Will ME, Suter GW. 1995. Toxicological benchmarks for screening potential contaminants of concern for effects on soil and litter invertebrates and heterotrophic process. Oak Ridge TN, USA: Oak Ridge National Laboratory. Report ES-ER-TM-126-R2.

CHAPTER 4

Soil Toxicity Tests—Invertebrates

Hans Løkke, Colin R. Janssen, Roman P. Lanno, Jörg Römbke, Sten Rundgren, Nico M. van Straalen .

Introduction

The soil-litter subsystem is one of the most complex components of the soil ecosystem in terms of taxonomic diversity and trophic relations, and metal pollution may affect soil ecosystems by altering the structure and function of soil invertebrate populations. Soil invertebrates occupy a central role in energy cycling pathways in soil systems as they represent many different feeding guilds (detritivores, herbivores, bacterial and fungal feeders, omnivores, and predators). Exposure of soil invertebrates to soil-associated contaminants may be through dermal uptake from the soil pore water and/or through ingestion of soil particles.

Usually, laboratory toxicity tests with soil organisms form the basis for the hazard and risk assessment of metals in soil ecosystems (EU 1996). The only soil invertebrate test formally accepted by the Organization for Economic Cooperation and Development (OECD) is the 14-day LC50 test using the earthworm *Eisenia fetida* (OECD 1984; USEPA 1996). A draft guideline also is available for a 28-day reproduction test using the springtail *Folsomia candida* (ISO 1999b). It has been stressed that the number of presently available "standard" test species is too limited to allow realistic hazard and risk assessment of soil-associated contaminants. Increasing the number of toxicity tests in a standard battery of tests by using soil invertebrates representing a wide range of life histories has been recommended by various authors (Van Straalen and Løkke 1997; Løkke and van Gestel 1998). Although short-term survival tests may provide valuable initial data on the ecotoxicity of soil contaminants, successful protection of soil invertebrate populations and the life-support function of soils requires information on effects of soil contaminants on reproduction of the species. As in other animal populations, long-term survival of soil invertebrates depends on reproductive capacity and recruitment of juveniles into the population.

The present chapter summarizes the views of a working group of soil invertebrate ecotoxicologists on the use and possible improvements of soil invertebrate tests for the hazard and risk assessment of metals and sparingly soluble metal compounds. Recommendations are made on how existing and newly developed test protocols

Test Methods to Determine Hazards of Sparingly Soluble Metal Compounds in Soils. Anne Fairbrother et al., editors.
©2002 Society of Environmental Toxicology and Chemistry (SETAC). ISBN 1-880611-42-2

may be modified for the evaluation of these types of substances. Extensive guidance is given on the selection of standard soil types for hazard identification and risk assessment, how soluble and sparingly soluble metal substances should be introduced into the test substrate, and how the tests should be conducted. Four soil invertebrate reproduction tests are suggested for use in the hazard assessments. Details on test design and performance are given, with special emphasis on toxicity evaluation of essential and nonessential metals. Guidance on quality assurance (QA) and quality control (QC) for the proposed procedures is suggested. Finally, a number of recommendations on research issues are made.

The Soil Substrate

The type of soil substrate selected for a test may have a profound influence on the test organism's response to added chemicals, both due to physico-chemical relationships between soil and chemical and to the organism's health within the test system (see Table 4-1). This section addresses these concerns and suggests appropriate soil matrices for conducting toxicity tests with metals and metal-containing substances.

Table 4-1 Possible candidates for test soils

Soil	Advantages	Disadvantages
Artificial according to OECD 1984	• Widely accepted in ecotoxicological effect testing • Standardized • Relatively easy to prepare • In accordance with hazard identification in the aquatic compartment: selection of one purely artificial test medium	• Not all important test organisms survive or reproduce in artificial soil (especially microorganisms) • Has a high amount of organic matter and a rare type of clay (kaolinite) • Currently, it is not possible to extrapolate results from such a soil to natural soils
Natural field	• Standard soils are available • Exposure situation is realistic (including bioavailability) • In accordance with similar considerations in the area of fate testing • Results also will be useful for other purposes such as environmental risk assessment	• Guidance is needed for which soil should be tested • The acceptable physiological range of soil physico-chemical characteristics for test species is not yet known

Selection of soil type

One of the key questions in soil invertebrate toxicity testing is what type of soil should be used in standard tests. The following criteria have to be considered when selecting soil for the purpose of hazard identification:

- preferably, the same test soils should be used for all testing of physico-chemical properties, and for toxicity testing of invertebrates, plants, and microorganisms;
- the number of soils should be kept low, but more than one soil is probably needed for 2 reasons:

  1) the chemistry of metals is complex and can be evaluated by use of varying soil properties and

  2) there are few soils in which all test species will survive and reproduce;

- the realistic worst-case situations should be represented;
- the data gained for hazard identification should, if possible, also be useful for environmental risk assessment; and
- the selection of soils should be, if possible, in agreement with new approaches such as the ecoregion concept (i.e., taking natural background levels of metals into account).

The workgroup considered the use of 5 so-called EUROSOILs, a group of well-characterized natural soils representative of European soil types and originally selected for fate testing; according to an OECD workshop held 1995 in Belgirate, Italy (Table 4-2). Standardized soils with similar physico-chemical characteristics can be found in North America and Japan (slight differences are tolerated). For example, EUROSOIL Number 3 corresponds to soil Number 1 in Japan, and

Table 4-2 Soil properties of EUROSOILs. Mean values of the reference sites are given. The range accepted by OECD is given in parenthesis (OECD 1995; Gawlik and Muntau 1999).

Nr	Reference site	Soil type and texture (FAO and U.S. system)	Organic carbon (%)	Clay (%)	pH* (KCl)
1	Aliminusa (Palermo) Sicily, Italy	Vertic cambisol, Brown Mediterranean Clay	1.3 (1 – 2)	75 (65 – 80)	5.1 (4.5 – 5.5%)
2	Souli (Corinth) Peloponnesos, Greece	Rendzina, Clay loam or loam	3.7 (3.5 – 6.5)	23 (20 – 40)	7.4 (> 7.5%)
3	Glamorgan Gwent (Cardiff) Wales, UK	Dystric cambisol, Acid, brown forest soil Loam	3.5 (3 – 4)	17 (15 – 30)	5.2 (4 – 5.5%)
4	Rots (Caen) Normandy, France	Orthic luvisol, Gray-brown podsolic silt loam	1.6 (1.5 – 3)	20 (15 – 25)	6.5 (5.5 – 7.5%)
5	Lauenburg, Schleswig-Holstein, Germany	Orthic podsol, Podsolized soils, Loamy sand or sand	10.3 (> 10)	6 (< 10)	3.2 (< 4.5%)

FAO = Food and Argiculture Organization
*Some species may not survive or reproduce in soils at pH 5 or lower. Acid soils simulate maximized bioavailability and a worst-case situation.

EUROSOIL Number 2 to soil number 14 in Japan (OECD 1995). Concerning the EUROSOILs, the above-mentioned disadvantages can be solved by 2 research initiatives: 1) short-term research: defining the ecological ranges of soil physico-chemical characteristics for test species and 2) long-term research: understanding the uptake mechanisms of metals for soil invertebrates.

The following soil properties also are available for these soils (measured at reference sites):

- grain size distribution, total carbon, $CaCO_3$, organic matter (OM), N, C/N ratio;
- organic S, P total, SiO_2, Al_2O_3, CaO, K_2O, Fe_2O_3, MgO, TiO_2;
- catonic exchange capacity (CEC), Fe total, Fe amorphous, Fe HCl-solution, Al amorphous, Al HCl-solution;
- mineralogical composition of clay and fine-silt (MI, Sm, Cl, Verm, Il, Kl, Q, F);
- Mn, Cu, Ca, K, Na, Mg, Zn, Cd, Pb, Ni, Cr, Sn, Ti, B, Sr, Mo, Co, Be, Ba, Li; and
- organochlorine compounds (i.e., DDT [dichlorodiphenyltrichloroethane], lindane, polychlorinated biphenyls, chlorinated benzenes).

Another possibility is the use of two soil types with the following characteristics that will represent scenarios of high metal availability for cationic and anionic metal-loids, respectively. These soils differ in key characteristics such as pH that determine chemical availability and probably other soil properties should be specified as well. Similar soil types with properties in the same ranges are found in North America and in other parts of the world.

For cationic metal species, test soils with the following properties should be used:

- pH 5 to 5.5,
- texture (loam or sandy loam),
- 2 to 3% OM, and
- low in Fe oxides.

This soil corresponds to the EUROSOIL Number 3.

For Cr(VI), As(V), U(VI), and similar metal species, test soils with the following properties should be used:

- pH 7.5 to 8,
- slightly calcareous is suitable,
- texture (loam or silt loam),
- 1 to 2% OM, and
- low in Fe oxides.

This soil corresponds to EUROSOIL Numbers 2 or 4.

The properties of soils that determine bioavailability are not completely understood at this time. It is not yet possible to extrapolate bioavailability from one soil to another by means of simple algorithms. Recently, Lock and Janssen (2001a), Lock et al. (2001), and Lock et al. (2000) have demonstrated that development of these types of simple, empirical models is possible and should be further investigated.

Incorporating test substances into the soil

The incorporation of metals and sparingly soluble metal compounds into the test soil is one of the most crucial aspects of the hazard identification of this type of substance. Metal salts, organometallics, and metallics should not be treated similarly. Organometallics should be treated as organic chemicals, although they usually are not dealt with in this context. If only one form of a metal is to be tested, a soluble salt is preferred.

Most test results described in the scientific literature have been derived from test soils spiked with readily soluble metal salts. This implies that observed biological responses were a function of both the metal and the anionic moiety that was added with the metal. The "salt effect" introduced in this way should not be ignored, especially when salt concentrations are high (Schrader et al. 1998). In some studies, these effects have been circumvented by supplementing all treatments in a concentration series with another soluble salt, so as to achieve an equal concentration of the anion over all metal concentrations and letting another cation, e.g., potassium, vary (Van Straalen et al. 1989). Another proposed method to remove excess anion and excess free metal is by percolating the soil with deionized water after spiking with a soluble metal salt (Smit and van Gestel 1998). However, these methods may be less appropriate when sparingly soluble metal compounds have to be spiked into test soils.

The invertebrate workgroup considered the following criteria for spiking methods:
- It should produce soils with different concentrations of the test metal in a reproducible manner in order to allow for data analysis in terms of concentration-response relationships.
- It should allow for homogeneous mixing of the metal compound through the soil to ensure that test organisms are exposed in a reproducible way.
- It should provide a certain "crumb structure" and density as suitable habitat for soil invertebrates.

The workgroup discussed several approaches that could be taken to address the spiking problem, recognizing that further research in this area is essential.

One approach could be to assume that the eventual distribution of metal species and adsorption states are independent of the route of entry into the soil and the compound introduced initially. One could then attempt to attain this distribution by artificial means following amendment of the soil with a soluble form of the metal (e.g., using a combination of accelerated aging, percolation, etc.). This type of "artificial and accelerated aging" of the soil has been investigated by Lock and Janssen (2002) using the artificial OECD soils and acute *Enchytraeus albidus* assays. This "soluble approach" does not really solve the problem of testing sparingly soluble metal compounds but circumvents it. The assumption of the ultimate state being independent of the metal compound introduced is questionable, at least within the time frame of interest in hazard identification. An advantage of this approach is that all compounds that share the same metal moiety are addressed in the same experiment.

Another approach would be to use the insoluble metal compound as it is, or as a powder with a known particle size distribution, but allow for a "transformation test" prior to introducing it into the soil. This could be done by shaking the compound with water for a defined period and by applying the resulting slurry to the soil.

A third possibility would be to mix the insoluble compound directly into the soil and allow the "transformation" to occur in the soil itself. Care would have to be taken to not destroy the structure of the soil to the point where it became unsuitable as a habitat for invertebrates. Disadvantages of this method are that the dissolution reactions in soil largely escape observation and that the necessary period for transformations to attain equilibrium is unknown (see also Chapter 2). It is expected that soil invertebrates themselves will accelerate the transformation process by increased mixing, gut passage, etc. Therefore, "engineering" invertebrates such as earthworms should be included in the soil during the transformation period.

Effects Assessment

This section describes the test species and test design most appropriate for hazard testing of metal-containing substances. Particular attention is paid to describing culture and test conditions that maintain the test organisms at optimum levels without nutrient-induced stress.

Test species

A single invertebrate test species does not adequately represent the vast number of invertebrate species in existence. Therefore, a battery of test organisms is proposed. Organisms should be representative of different

- taxonomical or physiological groups (e.g., Annelida, Arthropoda),
- ecological function (e.g., soil dwelling),
- trophic levels (e.g., saprophagous, predatory), and
- routes of exposure (e.g., via pore water, air, food).

Furthermore, the surface or volume ratio (size class: mesofauna and macrofauna) and possible "sinks" within the animal should be considered (e.g., sequestration of metals as seen in isopods and spiders) (Hopkin 1989). All of these factors require consideration, especially in higher-level testing required for risk assessment. Finally, only species that either have an internationally standardized and accepted test guideline or a draft version of such a guideline following OECD or the International Organization for Standardization (ISO) rules were selected.

Comparable proposals have been made for chemicals (Leon and van Gestel 1994; Samsøe-Petersen and Pedersen 1994; Koerdel et al. 1996; Römbke et al. 1996) and, very recently, soil contaminants in general (ISO 1999a).

Test design and performance

Range-finding tests

It is necessary to determine the range of concentrations to be applied in a definitive test. Therefore, a range-finding test must be conducted using 4 concentrations of the test substance in the range of either 1, 10, 100, and 1000 mg/kg (dry mass soil) for soluble compounds or 10, 100, 1000, and 10,000 mg/kg (dry mass soil) for sparingly soluble compounds. One replicate for each treatment plus a control is recommended. The test duration is usually 2 weeks (depending on the test system). The LC50 for mortality is estimated from the range-finding test. This value is used to determine the concentration range of the definitive test. As a first approximation, it is assumed that the EC10 of the definitive test is lower than the LC50 of the range-finding test by a factor of 10. However, it must be stressed that this is just an empirical relationship, which might be different in any given case (Crommentuijn et al. 1995). If no effects are observed in the range-finding test, even at the highest concentration of 1000 or 10,000 mg/kg, the definitive test can be designed as a limit test, comparing only 6 control vessels with 6 vessels containing soil with a concentration of 1000 or 10,000 mg/kg, depending on the test substance.

If available for a given test substance and/or soil, literature data can be used to define the range of concentrations for the definitive test in place of conducting a range-finding study.

Definitive tests

The number of replicates and test organisms depends on whether regression analysis or analysis of variance is chosen as the experimental design. The regression approach is a scientifically more reliable statistical design for the definitive test and follows recent statistical considerations (e.g., Pack 1993; Chapman et al. 1996; Joermann et al. 1996). In other words, we are not testing whether chemicals are toxic, but where (at what level) they become toxic. According to Paracelsus, a 16th century physician-alchemist, "All substances are poisons; there is none which is not a poison. The right dose differentiates a poison from a remedy," (Gallo 1996).

Practical reasons impose limits on replication and the number of concentrations that are feasible in the test. Therefore, in the definitive test, the following test design is recommended. A range of 12 concentrations spaced by a factor not exceeding 3 should be used. Two replicates for each treatment plus 6 controls are recommended, resulting in a total number of 30 test vessels. The spacing factor for the test concentrations may be variable, being smaller at low concentrations, and increasing at higher concentrations.

In the four tests recommended here, the endpoint is the number of juveniles at the end of the experimental period. As in the range-finding test, all other signs of toxic impact should be recorded. At first, the arithmetic mean number of juveniles and the associated variance per treatment or control for this endpoint are calculated.

For the purpose of hazard identification, the EC50 would be the most appropriate choice, since this is the most reliable value with the smallest variance. To compute any ECx value, the pretreatment means are used for regression analysis after an appropriate dose-response function has been found. An ECx is obtained by inserting a value corresponding to x% of the control mean into the equation found by regression analysis; 95% confidence limits are calculated according to Fieller (Finney 1971).

Treatment results also can be expressed as a percentage of the control result or as percent inhibition relative to controls, provided that control performance meets QA criteria. In these cases, probit or logistic sigmoid curves can be fitted to the results by means of a regression procedure. If hormetic effects are observed, analysis should be performed using a 4-parameter logistic or Weibull function, fitted by a nonlinear regression procedure.

If a limit test has been performed and the prerequisites (normality, homogeneity of variance) of parametric test procedures are fulfilled, the pairwise student-T-test may be used for means comparison. Otherwise the Mann-Whitney U-test procedure may be applied. However, these approaches based on NOECs (no-observed-effect concentrations) are no longer recommended and should be substituted by regression analysis.

Test duration

The advantages and disadvantages of acute and chronic tests and the possibility for developing a tiered test system were considered. Since all natural elements remain in the environment, reproduction or growth measurements are appropriate for hazard identification. Metal elements will remain in soils for a very long period, although bioavailability of metal ions will change over time. Therefore, it is recommended that long-term (reproduction) tests be conducted with all test species. Test duration will vary from 21 days (*Folsomia candida, Hypoaspis aculeifer*) to 8 weeks (*Eisenia andrei*). Extra work to be performed on reproduction testing as compared to counts of earthworm survival tests is minor. Survival test parameters (LC50) also may be obtained from reproduction range-finding tests.

Test conditions

Standard test conditions are given in available test protocols (confer with Table 4-3 and Appendix to Chapter 4). The most important parameters of the test substrate, such as food availability or soil properties (e.g., pH, OM content, texture) must be checked and, if necessary, adjusted according to the prescriptions in the testing

Table 4-3 Recommended invertebrate test species and guidelines (detailed in the Appendix to Chapter 4)

Test species	Guidelines
Annelids	
Eisenia andrei (Lumbricidae, earthworm)	ISO 1998
Enchytraeus albidus (Enchytraeidae, potworm)	ASTM 2000; OECD 2000; ISO 2001
Arthropods	
Folsomia candida (Collembola, springtail)	ISO 1999a
Hypoaspis aculeifer (Gamasina, predatory mite)	Krogh and Axelsen 1998

protocols. The physiological limits of the different test species should be documented to design the optimal test environment for control performance.

Ambient environmental conditions in the test chamber and test containers must be measured in all toxicity tests to ensure precise results. Environmental parameters to be monitored include temperature, soil water content (60 to 70% of water holding capacity), pH, and photoperiod. These will vary with the test species and must be maintained within a range that does not result in physiological stress. Guidance on the range for these conditions can be found in OECD (1995), Løkke and van Gestel (1998), ISO (1998), and Römbke and Moser (1999).

The moisture content in the soil toxicity test should be maintained at the level prescribed in the test protocol. This is accomplished through a reweighing and water addition procedure at selected intervals throughout the test. Some water should be allowed to evaporate by ventilating the headspace to prevent the growth of microbes.

Food is added in the proposed reproduction tests. Clean food should be analyzed for metal content before being added to the test system. Food should not be mixed into the soil (springtails will not eat it when prepared in this manner). Even though addition of food may affect metal bioavailability, the magnitude of this process is small and thus of minor importance to the final test result. The predatory mite (*Hypoaspis aculeifer*) test is an exception due to the fact that it feeds on living food items (e.g., *Folsomia candida*) moving around in the soil. Thus, mites will be exposed to metals taken up directly from the soil and to metals accumulated by their prey.

Background metal levels (Cu, Cd, Zn, Fe, Mn, Pb) in test soils should be measured and should be within a range that meets test organism requirements for essential metals and does not result in toxicity. Some test species may not be appropriate for conducting toxicity testing in some of the suggested test soils since normal soil conditions are outside their physiological tolerances. As an example, earthworm toxicity tests with *Eisenia andrei* should not be conducted in soils with a pH lower than 5.5. Measured soil parameters will determine that test species can be used in which soils. Research still is needed to clearly identify the range of environmental parameters required for the normal growth and reproduction of some suggested test species in the proposed soils.

Quality Assurance

Quality assurance and QC are necessary parts of any toxicity testing protocol. There are several specific QA issues related to toxicity tests conducted with metals and sparingly soluble metal compounds, but in all tests, a detailed documentation of any procedure, operation, or result is required. Since the results of toxicity tests frequently are used for regulatory purposes, most countries will require strict adherence to a certain level of QA. This might include a good laboratory practice (GLP) certification of the testing facility, preparation of standard operation procedures (SOPs), and yearly testing of unknown samples. The details of these requirements are understood by individual laboratories and will not be discussed here. The discussion that follows will be limited to QA and QC areas specifically related to conducting soil invertebrate toxicity tests in soils for the purposes of hazard identification for metals and sparingly soluble metal compounds.

Characterization of test conditions

Specific chemical and physical properties of the test soil and the test conditions must be documented prior to commencement of any test. These include temperature; soil water content and water holding capacity; pH; photoperiod; texture; OM; CEC; and total Ca-hydroxides, Fe-hydroxides, and Mn-hydroxides. Moisture content and pH (1 M KCl) must also be determined, at a minimum, at the beginning and end of each test. During longer tests, such as the earthworm reproduction tests, moisture content should be determined gravimetrically every week, and if moisture content decreases, deionized water should be added to replenish moisture content.

Total metal concentrations must be determined in the culturing medium, in replicate and control test soils, and in food offered to the test organisms. At the present time, no ranges of acceptable background values for metals in these media have been defined, but they should be documented. Metal levels in the culturing medium and test soil should be maintained above a requirement level for essential elements (e.g., Zn, Cu, Fe, Co, Mn, Se) and below a level that will select for metal-tolerant genotypes in the population of test organisms in use. Background soil levels for the region where the test soil was collected usually will approximate these requirements.

All measurements of soil properties and metal concentrations in the culture medium, food, and test soil must follow internationally standardized analytical chemistry guidelines (preferably ISO).

Culturing procedures

One of the most important aspects of assuring consistent, precise responses of soil organisms during toxicity tests with metals is ensuring that previous contact with elevated levels of metals has been eliminated. It is recognized that increased metal tolerance can occur either through physiological acclimation or genetic adaptation if test organisms are exposed to elevated levels of metals (Reinecke et al. 1999). Although the concept of inducing metal tolerance by metal preexposure has been documented (Hopkin 1989; Posthuma and van Straalen 1993), further research is required to establish the levels of metals in the soil and diet of test organisms that do not induce tolerance. It is suggested that a screening procedure be established to ensure that physiological acclimation or genetic adaptation to metals does not occur in the culture systems. The origin of test organisms used during the toxicity test must be documented and it must be ensured that organisms have no history of prior exposure to elevated levels of heavy metals, either through contact with metal-contaminated soil or the ingestion of metal-contaminated food.

Reference toxicity tests should be conducted biannually with organisms representative of laboratory cultures or other test organism sources. These tests will be conducted with a positive control compound (carbendazim for *Eisenia andrei* and *Enchytraeus albidus*, dimethoate for *Folsomia candida* and *Hypoaspis aculeifer*) and will ensure precision in the sensitivity of the test species during toxicity tests. Control charts should be constructed to monitor reference toxicity test results over time. Reference chemicals should be tested in actual test soil used in the definitive test according to the requirements specified in the individual protocols (Løkke and van Gestel 1998).

Validity criteria

Test results will be considered valid if conditions described in the individual test guidelines (Løkke and van Gestel 1998) are met. These include parameters such as average control mortality at the end of range-finding and definitive tests and the ratio of the number of juvenile organisms per test vessel at the end of the definitive test to the number of adult animals that were introduced into the test vessels at the beginning of the test.

Test report

Test reports shall refer to the standard protocols and must contain a summary of the methods used, parameters measured, and results obtained during the study. The test report shall provide at least the following information:

- A full description of the experimental design and procedures, including a description of the soil and test equipment used;
- Chemical identification of the test substance according to International Union of Physicists and Chemists (IUPAC) nomenclature, batch, lot, and Chemical Abstracts Service (CAS) -number, structural formula, and purity of the test substance;
- Properties of the test and reference substance;
- Method of application of the substance to the soil;
- Measured concentrations of test substances in all replicate soils;
- Identification of the test organism and description of stock cultures;
- Description of the culturing conditions;
- Source of supply of the test organism;
- Description of the test conditions, including moisture content and pH of the soil at the start and end of the test;
- Mortality of the adults and the number of juveniles at the end of the range-finding test or at the end of the definitive test (depending on the test system);
- Description of obvious physical or pathological symptoms or distinct changes in behavior observed in the test organisms;
- Statistically calculated values (LC50, ECx) using measured total metal concentrations, including 95% confidence limits, method of calculation, and plots of dose-response relationships;
- All information, including all measured raw data, developed during all phases of testing with the test and reference substance;
- Discussion of the results.

Other Issues

The testing strategy described above was developed for hazard identification of metal compounds. Nevertheless, it also can be considered as the first step in a tiered approach for risk assessment. Results obtained from these tests can be used in conjunction with exposure assessments, to provide an initial characterization of risks in the context of regional risk assessment and site-specific risk assessment. However, it should be recognized that the species chosen for testing do not cover the rich biodiversity of invertebrate life in soil. According to the Convention on Biological Diversity agreed on at the Earth Summit in Rio de Janeiro 1992 (Glowka et al. 1994), biodiversity includes 3 levels of variability:

1) genetic variability within a species,

2) species richness of communities, and

3) diversity of ecosystems and landscapes.

A complete assessment of effects of metals on biodiversity of the soil invertebrate community is outside the scope of the present document.

Soil invertebrates often are used in combination with microorganisms for multiple species and microcosm testing. Such tests presently are being standardized and developed further, e.g., terrestrial model ecosystems (Morgan and Knacker 1994; Sheppard 1997). Other approaches in this field are the use of litterbags to study decomposition of OM (Kula and Römbke 1998), and the use of bait laminas (Kratz 1998) to study feeding activity of invertebrates. There certainly is a need for functional testing for invertebrates, but such tests fall outside the scope of hazard identification.

Concentrations of metals inside the bodies of invertebrates are often considered indicators of effects. Bioaccumulation of metals was discussed at the technical workshop, Test Methods to Determine Hazards of Sparingly Soluble Metal Compounds in Soils, held in 1995 under the auspices of the Canada/European Union (EU) Metals and Minerals working group in Brussels, Belgium (Anonymous 1996). For essential metals, measurement of total internal concentrations in invertebrates is of limited value for hazard and risk assessment because this concentration is usually regulated within a narrow range over a broad range of environmental concentrations (Lock and Janssen 2001b). Rather than the internal concentration, the rate of entry of metals into the body can indicate risk, however, this is difficult to measure. For nonessential elements, internal concentrations can be useful indicators of risk, provided that they can be related to a critical internal threshold such as the lethal body concentration (Van Straalen 1996). This information is available for only a few invertebrate species. To use internal concentrations meaningfully, knowledge about the toxicokinetics (uptake and excretion rates) also is essential (Lanno et al. 1998). Bioaccumulation is considered a suitable approach for assessment of bioavailability in the field and transfer of metals in the food chain, however, it falls outside the scope of hazard identification. In the field of risk assessment, transfer of metals to invertebrate-feeding higher animals (e.g., moles, shrews, thrushes) is an important issue. Critical concentrations in target organs of small predatory mammals have been suggested for use in setting criteria for metals in soil (Ma and van der Voet 1993).

Recommendations

The workgroup identified several areas where further research is required and made a series of recommendations.

The following are recommendations that have special relevance to hazard identification:

- It is proposed to classify hazards of sparingly soluble metal compounds to soil invertebrates on the basis of reproduction toxicity tests with four different test species as described above.
- There is an urgent need to determine the ecological requirements of the test species in the selected test soils, including the effect ranges of the reference compounds in these soils.
- Research is needed to develop new methods for incorporating sparingly soluble metal compounds into the soil.
- It is recognized that increased metal tolerance can occur either through physiological acclimation or through genetic adaptation. It is suggested that a screening procedure be established to ensure that acclimation or adaptation does not occur in the culture systems.
- Research is required to establish physiological ranges for essential metals in soil organisms that do not induce deficiency syndromes.

Recommendations that are relevant to risk assessment include the following:

- Soil properties are highly variable. It is difficult to derive simple model relationships between soils and bioavailability of different metals. However, it is recommended that algorithms be developed to make comparisons of toxic responses in different soils possible.
- There is a need for development of tests that assess the function of invertebrates in soil ecosystems. An example is the use of bait lamina that measures feeding activity of soil invertebrates in semi-field and field studies.
- Effects on mammals (e.g., moles, shrews) and birds (e.g., thrushes) that feed intensively on soil invertebrates and may accumulate metals should be investigated further and be included in risk assessments of contaminated soils.

APPENDIX TO CHAPTER 4:

Description of the Four Recommended Tests

Table 4-4 Reproduction and growth test with the earthworm *Eisenia fetida*

Principle	Sublethal laboratory test using earthworms
Guideline	BBA 1994; related guidelines show only slight modifications (e.g., ISO 1998; Kula and Larink 1998)
Test species	*Eisenia fetida* (Savigny 1826) or *E. andrei* (Andrei 1963); compost worm (Lumbricidae); synchronized laboratory mass culture
Test design	10 worms per test vessel (e.g., Bellaplast container); 2 to 12 months old (250 to 600 mg fw); acclimatization period: 1 to 7 days
Substrate	Artificial soil according to OECD 1984: quartz sand, kaolin clay, *Sphagnum* peat, calcium carbonate, and water (or standard field soil lufa 2.2)
Feeding	Finely ground cow manure, 5 g per test vessel mixed into the soil at the beginning of the test and after removal of the adults; in between weekly feeding of 5 g per test vessel (if the food is not consumed, reduction of amount is necessary)
Parameter	Mortality, biomass, morphological, and behavioral changes of the adult worms; number of juveniles at the end of the test
Duration	2 months; examinations 0, 28, and 56 days after application
Application	Spraying of the test substance on top of the test substrate (including worms) using a laboratory spraying advice (BBA 1994) or by mixing (ISO 1998)
Concentration	Pesticides: 1x and 5x of the maximum recommended application rate; other chemicals. Range-finding test: 0, 1, 10, 100, and 1000 mg/kg dw (one replicate); Definitive test: several concentrations (spacing factor < 2) (4 replicates); not higher than 1000 mg/kg dw
Performance	Temperature: 20 ± 1 °C: light-dark cycle 8:16, 12:12 or 16:8 hours at 400 to 800 lux; feeding with finely ground cattle manure; extraction of the juveniles by means of water bath, sieving, or via hand-sorting
Reference substance	BBA 1994: Carbendazim or benomyl (concentration according to 750 g a.i./ha); ISO 1998: LOEC of carbendazim (1 to 5 mg a.i./kg soil dw)
Validity criteria	Control: mortality after 14 days ≤ 10%; decrease of biomass ≤ 20%; at least 30 juveniles per test vessel at the end of the test; coefficient of variation of average juvenile numbers < 50% (BBA 1994) or 30% (ISO 1998)
Assessment	Evaluation with "suitable" statistical methods, preferably regression analysis
Notes	A detailed description of the test development and the results of two ring tests are given in Riepert and Kula 1996; in the ISO guideline, adaptations for soil quality assessment are described

a.i. = active ingredient
dw = dry weight
fw = fresh weight

Table 4-5 Enchytraeid reproduction test

Principle	Sublethal laboratory test using saprophagous annelids (Enchytraeidae)
Guideline	ASTM 2000; OECD 2000; ISO 2001
Test species	*Enchytraeus albidus*; *Enchytraeus sp.* (Enchytraeidae)
Test design	10 adult worms (as identified by visible eggs in the clitellum region) per test vessel (e.g., 0.2 to 0.25 L glass with lid)
Substrate	Artificial soil according to OECD 1984: quartz sand, kaolin clay, *Sphagnum* peat, calcium carbonate, and water
Feeding	Finely ground rolled oats, 50 mg per test vessel at the beginning of the test; afterwards, except at day 28, weekly feeding of 25 mg per test vessel
Parameter	Range-finding test: mortality, behavior
	Definitive test: Number of juveniles
Duration	Range-finding test: 2 weeks
	Definitive test: 6 weeks: Removal of the adult worms after 3 weeks; counting of the juveniles hatched after 3 more weeks (for faster reproducing *Enchytraeus sp.*, test duration can be reduced to 3 weeks)
Application	Mixing of the test substance in the test substrate (without worms); if not water-soluble, mixing with quartz sand
Concentration	Range-finding test: Control 0, 0.1, 1, 10, 100, 1000 mg/kg (1 replicate)
	Definitive test, NOEC design: 5 concentrations plus control (4 replicates) (spacing factor 1.8) ECx design: 12 concentrations (2 replicates)
Performance	Room temperature: 20 ± 2 °C; weekly feeding (oats strewn on soil surface); permanent light; Moisture: 40 to 60% of the WHC_{max}; Counting of the hatched juveniles after staining with Bengal red
Reference substance	EC50 (reproduction) for carbendazim: 1.2 ± 0.8 mg a.i./kg dw
Validity criteria	Control: Mortality < 10% after 3 weeks; number of juveniles per test vessel > 25 after 6 weeks; Coefficient of variance (reproduction) less than 50%.
Assessment	Determination of the NOEC by ANOVA or (preferably) determination of the ECx using regression analysis
Notes	Recently the test was validated in an international ring test; performance in field soils for soil quality assessment is possible

a.i. = active ingredient
ANOVA = analysis of variance
dw = dry weight
NOEC = no-observed-effect concentration
WHC = water holding capacity

Table 4-6 Reproduction toxicity test with collembola

Principle	Sublethal laboratory test using one springtail species
Guideline	ISO 1999a based on Riepert and Kula 1996
Test species	*Folsomia candida* (Willem 1902) (Isotomidae), springtail; synchronized laboratory mass culture
Test design	10 individuals (10 to 12 days old) per test vessel (100 mL glass vessels with lid [diameter 5 cm]), filled with 30 g fw of test substrate
Substrate	Artificial soil according to OECD 1984: quartz sand, kaolin clay, *Sphagnum* peat, calcium carbonate, and water
Feeding	Granulated dry yeast, 2 mg per test vessel at the beginning of the test and after 14 days (if the food is not consumed, reduction of amount is necessary)
Parameter	Mortality and reproduction
Duration	Examination after 4 weeks
Application	According to the solubility of the test substance: dissolved in water or in an organic solvent in a mixture with finely ground quartz sand; if insoluble in water and organic solvents, mixed with 10 finely ground quartz sand
Concentration	Range-finding test (optional): 4 concentrations (e.g., 1, 10, 100, 1000 mg/kg) and a control Definitive test (not higher than 1000 mg/kg dw), NOEC approach: At least 5 concentrations, organized in a geometrical series (spacing factor not exceeding 2), 5 replicates ECx approach: 12 concentrations, spacing factor might be variable, 2 replicates (control: 5 replicates)
Performance	Temperature: 20 ± 2 °C; light-dark cycle of 12:12 hours or 16:8 hours with 400 to 800 lux; Moisture: 40 to 60% WHC_{max} (compensation necessary if the loss exceeds 2%); feeding with 2 mg of granulated dry yeast; determination of springtails (adults and juveniles) by adding water to the test vessels and counting of the animals floating at the surface (e.g., by using a counting grid or on a projected slide)
Reference substance	LOECs: E 605 forte (a.i. Parathion 507.5 g/L): 0.10 to 0.32 mg/kg dw or Betanal plus (a.i. phenmedipharm 160 g/L): 100 to 200 mg/kg
Validity criteria	Control: mortality $\leq$ 20%; minimal reproduction of 100 juveniles per replicate; CV (reproduction) $\leq$ 30%
Assessment	NOEC, LOEC, and ECx by using "suitable" statistical methods (e.g., multiple T-test and, preferably, regression analysis)
Notes	The history and the validation of this test is documented in detail in Riepert and Kula 1996 (It also can be used in soil quality assessment)

a.i. = active ingredient
CV = coefficient of variation
dw = dry weight
fw = fresh weight
LOEC = lowest-observed-effect concentration
NOEC = no-observed-effect concentration
WHC = water holding capacity

Table 4-7 Sublethal effects on predatory mites and springtails

Principle	Sublethal laboratory test using a two-species system: Predatory mite and springtails serving as food
Guideline	Formalized guideline proposal (Krogh and Axelsen 1998)
Test species	*Hypoaspis aculeifer* Canestrini (Gamasina), predatory mite; *Folsomia fimetaria* L. (Isotomidae), springtail; synchronized laboratory culture
Test design	10 male and 5 female adults of *H. aculeifer* and 100 individuals of *F. fimetaria* (16 to 19 days old at the start of the test) per test vessel (glass cylinder: 6 cm diameter, 5.5 cm high)
Substrate	Artificial soil according to OECD 1984: Quartz sand, kaolin clay, *Sphagnum* peat, calcium carbonate, and water or lufa standard field soil 2.2; each with 50% WHC_{max} (modifications according to the requirements of the animals are possible); in total an amount of 60 g fw
Feeding of prey	Granulated dry yeast, 15 mg per test vessel at the beginning of the test and after 14 days (if the food is not consumed, reduction of amount is necessary)
Parameter	Predatory mites: mortality and growth of the adults and the number of juveniles (reproduction)
Duration	Examination after 3 weeks
Application	According to the solubility of the test substance: Dissolved in water or in an organic solvent in a mixture with finely ground quartz sand; if insoluble in water and organic solvents, mixed with 10 finely ground quartz sand
Concentration	Range-finding test: 0.1, 1, 10, 100, 1000 mg/kg dw; Definitive test: Not specified number of concentrations for determination of an ECx but not higher than 1000 mg/kg dw; at least 4 replicates
Performance	Temperature: 20 ± 1 °C; light-dark cycle: 12:12 hour with 400 to 800 lux; feeding of the collembolans with 15 mg baker's yeast at day 0 and after 14 days; the mites fed on the springtails; dry extraction method; counting either manually under a stereo microscope or by means of digital image processing
Reference substance	Insecticide dimethoate: EC50 (reproduction of the mites) 2.0 to 3.0 mg/kg dw; Standard field soil lufa 2.2 (at approximately 1 mg/kg dw; a hormesis effect is sometimes observed)
Validity criteria	Control: Mortality of the female mites not higher than 10% and at least 20 juvenile mites per test vessel at the end of the test
Assessment	EC50, EC10, and LC50 determined by using "suitable" statistical methods, preferably regression analysis
Notes	One of the few examples of a two-species test laboratory system

dw = dry weight
fw = fresh weight
WHC = water holding capacity

References

Anonymous. 1996. Canada/European Union Technical Workshop on Biodegradation/Persistence and Bioaccumulation/Biomagnification of Metals and Metal Compounds; 11-13 December 1995. Brussels, Belgium: Canada/European Union Metals and Minerals working group. 48 p.

[ASTM] American Society for Testing and Materials. 2000. Standard guide for conducting laboratory soil toxicity or bioaccumulation tests with the lumbricid earthworm *Eisenia fetida* and the enchytraeid potworm *Enchytraeus albidus*. In: Annual book of standards. Conshohocken PA, USA: ASTM. Guideline nr E 1676-97.

[BBA] Biologische Bundesanstalt für Land- und Forstwirtschaft. 1994. Richtlinienvorschlag für die Prüfung von Pflanzenschutzmitteln (nr VI, 2-2): Auswirkungen von Pflanzenschutzmitteln auf die Reproduktion und das Wachstum von *Eisenia fetida/Eisenia andrei*. Braunschweig, Germany: BBA. 11 p.

Chapman PM, Caldwell RS, Chapmann PF. 1996. A warning: NOECs are inappropriate for regulatory use. *Environ Toxicol Chem* 15:77-79.

Crommentuijn T, Doodeman CJAM, van der Pol JJC, Doornekamp A, Rademker MCJ, van Gestel CAM. 1995. Sublethal sensitivity index as an ecotoxicity parameter measuring energy allocation under toxicant stress: Application to cadmium in soil arthropods. *Ecotoxicol Environ Saf* 31:192-200.

[EU] European Union. 1996. Technical guidance documents in support of Directive 93/67/EEC on risk assessment of new notified substances and regulation (EC) nr 1488/94 on risk assessment of existing substances. Part II. Brussels, Belgium: EU. p 241-503.

Finney DJ. 1971. Probit analysis. 3rd ed. Cambridge, UK: Cambridge University Press. p 19-76.

Gallo MA. 1996. History and scope of toxicology. In: Klaassen CD, editor. Casarett and doull's toxicology: The basic science of poisons. 5th ed. New York NY, USA: McGraw-Hill. p 3-11.

Gawlik BM, Muntau H, editors. 1999. Eurosoils II. Laboratory reference materials for soil-related studies. Luxembourg, Germany: Office for Official Publications of the European Communities, Joint Research Centre. EUR 18983.

Glowka L, Burhenne-Guilmin F, Synge H, McNeely JA, Gundling L. 1994. A guide to the convention on biological diversity. Cambridge, UK: IUCN. 161 p.

Hopkin SP. 1989. Ecophysiology of metals in terrestrial invertebrates. London, UK: Elsevier Applied Science Publication, Ltd. 366 p.

[ISO] International Organization for Standardization. 1998. Soil quality—Effects of pollutants on earthworms (*Eisenia fetida*). Part 2: Determination of effects on reproduction. Geneva, Switzerland: ISO. ISO 11268-2.

[ISO] International Organization for Standardization. 1999a. Soil quality—Guidance on the ecotoxicological characterization of soils and soil materials. Geneva, Switzerland: ISO. ISO/DIS 15799.

[ISO] International Organization for Standardization. 1999b. Soil quality—Inhibition of reproduction of Collembola (*Folsomia candida*) by soil pollutants. Geneva, Switzerland: ISO. ISO 11267.

[ISO] International Organization for Standardization. 2001. Soil quality—Effects of pollutants on Enchytraeidae (*Enchytraeus* sp.). Determination of effects on reproduction. Geneva, Switzerland: ISO. CD 16387.

Joermann G, Köpp H, Kula C. 1996. Fachgespräche zur Statistik in der Ökotoxikologie. Heft 17. Braunschweig, Germany: BBA.

Koerdel W, Hund K, Klein W. 1996. Erfassung und Bewertung stofflicher Bodenbelastungen. UWSF-Z. *Umweltchem Ökotox* 8:97-103.

Kratz W. 1998. The bait-lamina test – General aspects, applications and perspectives. *Environ Sci Pollut Res* 5:94-96.

Krogh PH, Axelsen JA.1998. Test on the predatory mite *Hypoaspis aculeifer* preying on the collembolan *Folsomia fimetaria*. In: Løkke H, van Gestel CAM, editors. Handbook of soil invertebrate toxicity testing. London, UK: John Wiley and Sons, Inc. p 239-251.

Kula C, Römbke J. 1998. Testing organic matter decomposition within risk assessment of plant protection products. *Environ Sci Pollut Res* 5:55-60.

Kula H, Larink O. 1998. Tests on the earthworms *Eisenia fetida* and *Aporrectodea caliginosa*. In: Løkke H, van Gestel CAM, editors. Handbook of soil invertebrate toxicity testing. London, UK: John Wiley and Sons, Inc. p 95-112.

Lanno RP, LeBlanc SC, Knight BL, Tymowski R, Fitzgerald DG. 1998. Application of body residues as a tool in the assessment of soil toxicity. In: Sheppard, Bembridge, Holmstrup, Posthuma, editors. Advances in earthworm ecotoxicology. Pensacola FL, USA: SETAC. p 41-53.

Leon CD, van Gestel CAM. 1994. Selection of a set of laboratory ecotoxicity tests for the effects assessment of chemicals in terrestrial ecosystems. Paris, France: OECD. Discussion Paper.

Lock K, Janssen CR. 2001a. Modelling zinc toxicity for terrestrial invertebrates. *Environ Toxicol Chem* 20:1901-1908.

Lock K, Janssen CR. 2001b. Zinc and cadmium body burdens in terrestrial invertebrates: Use and significance in environmental risk assessment. *Environ Toxicol Chem* 20:2067-2072.

Lock K, Janssen CR. 2002. The effect of ageing on the toxicity of zinc for the potworm *Enchytraeus albidus*. *Environ Pollut* 116:289-292.

Lock K, Janssen CR, De Coen WM. 2000. Multivariate test designs to assess the influence of zinc and cadmium bioavailability in soils on the toxicity to *Enchytraeus albidus*. *Environ Toxicol Chem* 19:2666-2671.

Lock K, Janssen CR, De Coen WM. 2001. Test designs to assess the influence of soil characteristics on the toxicity of copper and lead to the oligochaete *Enchytraeus albidus*. *Ecotoxicology* 10:137-144.

Løkke H, van Gestel CAM, editors. 1998. Handbook of soil invertebrate toxicity testing. London, UK: John Wiley and Sons, Inc. 281 p.

Ma W-C, van der Voet H. 1993. A risk assessment model for toxic exposure of small mammalian carnivores to cadmium in contaminated natural environments. *Sci Total Environ* (Suppl):1701-1714.

Morgan E, Knacker T. 1994. The role of laboratory terrestrial model ecosystems in the testing of potentially harmful substances. *Ecotoxicology* 3:213-233.

[OECD] Organization for Economic Cooperation and Development. 1984. OECD guideline for testing of chemicals. Earthworm acute toxicity test. Paris, France: OECD. Guideline nr 207.

[OECD] Organization for Economic Cooperation and Development. 1995. Final report of the OECD Workshop on Selection of Soils/Sediments; Belgirate, Italy. Paris, France: OECD.

[OECD] Organization for Economic Cooperation and Development. 2000. OECD guideline for testing of chemicals. Enchytraeidae reproduction test. Paris, France: OECD. Guideline nr 220 (draft).

Pack S. 1993. A review of statistical data analysis and experimental design in OECD aquatic toxicology test guidelines. Sittingbourne, UK: Shell Research Ltd. 42 p.

Posthuma L, van Straalen NM. 1993. Heavy-metal adaptation in terrestrial invertebrates: A review of occurrence, genetics, physiology and ecological consequences. *Comp Biochem Physiol* 106C:11-38.

Reinecke SA, Prinsloo MW, Reinecke AJ. 1999. Resistance of *Eisenia fetida* (oligochaeta) to cadmium after long-term exposure. *Ecotoxicol Environ Saf* 42:75-80.

Riepert F, Kula C. 1996. Development of laboratory methods for testing effects of chemicals and pesticides on collembola and earthworms. Parey Buchverlag, Berlin, Germany: Mitteilungen aus der Biologischen Bundesanstalt für Land- und Forstwirtschaft. 82 p. Heft 320.

Römbke J, Bauer C, Marschner A. 1996. Hazard assessment of chemicals in soil. Proposed ecotoxicological test strategy. *Environ Sci Pollut Res* 3:78-82.

Römbke J, Moser T. 1999. Organization and performance on an international ringtest for the validation of the enchytraeid reproduction test. Volume I and II. Berlin, Germany: Umweltbundesamt. UBA-Text 4/99. 150 p, 233 p.

Samsøe-Petersen L, Pedersen F. 1994. Discussion paper regarding guidance for terrestrial effects assessment. Horsholm, Denmark: VKI, Water Quality Institute. 48 p.

Schrader G, Metge K, Bahadir M. 1998. Importance of salt ions in ecotoxicological tests with soil arthropods. *Appl Soil Ecol* 7:189-193.

Sheppard SC. 1997. Toxicity testing using microcosms. In: Tarradellas J, Bitton G, Rossel D, editors. Soil ecotoxicology. Boca Raton FL, USA: Lewis Publishers. p 345-373.

Smit CE, van Gestel CAM. 1998. Effects of soil type, prepercolation, and aging on bioaccumulation and toxicity of zinc for the springtail *Folsomia candida*. *Environ Toxicol Chem* 17:1132-1141.

[USEPA] U.S. Environmental Protection Agency. 1996. Earthworm subchronic toxicity test. USEPA. OPPTS nr 850.6200. Report nr 712-C-96-167.

Van Straalen NM. 1996. Critical body concentrations: Their use in bioindication. In: Van Straalen NM, Krivolutsky DA, editors. Bioindicator systems for soil pollution. Dordrecht, Netherlands: Kluwer Academic Publisher. NATO ASI Series 2: Environment 16:5-16.

Van Straalen NM, Schobben JM, De Goede RGM. 1989. Population consequences of cadmium toxicity in soil microarthropods. *Ecotoxicol Environ Saf* 17:190-204.

Van Straalen NM, Løkke H. 1997. Ecological risk assessment of contaminants in soil. London, UK: Chapman and Hall. 333 p.

CHAPTER 5

Terrestrial Plant Toxicity Tests

Frank van Assche, Jose L. Alonso, Lawrence A. Kapustka, Richard Petrie,
Gladys L. Stephenson, Robert Tossell

Introduction

Hazard assessment of metals and metalloid compounds in the terrestrial environment includes the characterization of toxic effects, if any, on plants. However, there are several technical challenges for plant toxicity testing, particularly for sparingly soluble compounds. These generally relate to designing standard and site-specific (framework) toxicity test procedures, which can be used to compare hazard assessment data worldwide. The test procedures should be relatively easy to administer, minimize variability, and have high precision. Key issues in terrestrial plant toxicity testing are identified and discussed in this chapter. Requirements specific to metals and their compounds are addressed in the recommendations. Existing standard test guidelines (ASTM 1999; OECD 2000) and regulatory processes in the European Union (EU) and North America, for which hazard assessment data are used, framed the test recommendations presented here.

Regulatory framework

Hazard assessment forms the basis for a number of regulatory processes, including:

- hazard assessment of substances for effects on terrestrial plants,
- evaluation of phytotoxic effects of metals for regional and site-specific environmental risk assessments, and
- evaluation of phytotoxic effects of metal compounds and remediation of sites contaminated with these substances.

The requirements for testing differ among these 3 assessment frameworks, as well as the regulatory environment for which the data are used (Table 5-1). The primary focus is on the testing requirements for hazard assessment, though consideration of needs for risk assessment and site characterization are included. Recommendations for further research to improve precision and consistency among laboratories are presented.

Test Methods to Determine Hazards of Sparingly Soluble Metal Compounds in Soils. Anne Fairbrother et al., editors.
©2002 Society of Environmental Toxicology and Chemistry (SETAC). ISBN 1-880611-42-2

Table 5-1 Use and requirements for hazard assessment in different regulatory processes

	Classification for soil effects (EU-OECD)	Environmental risk assessment (EU-OECD)	Contaminated sites: evaluation or remediation
Required test	Standard	Standard + nonstandard	Standard + nonstandard
Required result	1 (lowest) toxicity value	PNEC derived from whole database	Toxicity screening
Environmental application	1 generic environment*	Local or regional scale	Local scale (site-specific)

OECD = Organization for Economic Cooperation and Development
PNEC = predicted no-effect concentration (EU risk assessment)
*Ecological entities distinguished in ecological risk assessment of natural elements are considered areas characterized by a distinct availability of background metal levels to biota. Biota living in such distinct areas are considered to be conditioned to natural background metal levels and therefore behave differently in ecotoxicity tests. This concept has recently been developed for the freshwater environment ("metalloregions") and probably is more applicable to the terrestrial environment, as background metal levels and their availability to biota can differ quite markedly among areas.

Hazard classification

Regulatory guidelines for hazard classification (e.g., EU guidance) handle acute and chronic tests differently. Test results from acute exposures are used to evaluate potential for immediate effects; results from chronic exposures are used to assess longer-term consequences. Explicit distinction in the guidance makes it is necessary to distinguish clearly the differences between acute and chronic plant tests. However, there is no consensus among plant toxicologists regarding such distinctions. Clearly, there is agreement that very short exposure periods (minutes or hours in duration) are acute exposures and that exposures that span the expected life of a plant are chronic exposures. Phytotoxicity tests that measure only germination percentage or germination rates are considered to be acute tests. Life-cycle tests such as the Brassica life-cycle test are clearly classified as chronic exposures. Most other standardized tests fall between these extremes, with the designation of "acute" or "chronic" being an arbitrary distinction. Operational definitions are required. For example, chronic exposures for plants are those that span a majority of the life expectancy of the species or a significant portion of the vegetative and reproductive life stages. Endpoints related to growth and biomass can be considered in both acute and chronic exposure scenarios. Reproductive endpoints, measured in chronic exposure tests, are often preferred, but end users should be aware that plants often allocate more energy to fruit or seed production when stressed than is the case for normal, nonstressed plants. Also, reproductive endpoints may be the primary measurements taken following onetime acute exposures such as an application of herbicides. The large differences among plants regarding time to complete phenological stages add to the difficulty of defining exposure categories as acute versus chronic. Some short-lived annuals germinate within 1 to 2 days and can complete their life cycle within 4 to 6 weeks. Long-lived perennials may require months for seeds to germinate, decades to reach sexual maturity, and centuries before normal death.

The operational definitions used in this chapter are predicated on testing conditions in which metals or metal-containing substances are incorporated into a soil matrix, exposure to plants occurs solely through the roots (or root-mycorrhizal complex), and plant responses are evaluated after some period of exposure. Secondly, the definitions apply to herbaceous plants, which propagate primarily by seed, have relatively high rates of germination over a period of few days, and have no unusual germination requirements (e.g., stratification, scarification)[1]. Germination tests and seedling emergence tests (14 days post emergence) are considered to be acute tests. Longer exposure periods might provide a better opportunity for expression of effects and there is a greater likelihood of detecting differences among treatments. However, an extended exposure period (i.e., 21 days, 28 days, or 56 days) may necessitate the use of nutrient amendments or greater soil volume in each test unit. The confounding effects of nutrients added to a test soil may confound toxic effects in some cases, requiring substantially greater complexity of the experimental design (e.g., a factorial design with varying nutrients arrayed over the test substance concentrations). Therefore, it is recommended that a 14-day exposure duration be used for the purposes of hazard classification. Longer exposure periods may be appropriate for risk assessment; it depends on the objectives of the risk assessment. Contaminated site assessments and phytoremediation might require different species and a determination of tolerance rather than sensitivity, but the basic testing procedures would be the same.

Risk assessment

Several measurement endpoints (discussed in detail in section Species and measurement endpoint selection) may be relevant for assessing risks of metal substances to plants. No preference is given to tests of short or long duration. There is no information, to the best of our knowledge, indicating that reproductive endpoints measured in chronic exposure studies are more sensitive to metals than are vegetative endpoints measured in acute exposures.

Significant refinements to generic hazard assessment values are required for ecological risk assessments (ERAs). The exposure term for a risk equation might be estimated from algorithms that describe bioavailability and flux rates but to date, no such algorithms exist. Alternatively, ecological data generated from representative soils may be used to generate regional or site-specific determinations of phytotoxic responses. Risk characterization may incorporate spatial and temporal patterns of the bioavailable fractions or surrogate values (e.g., pH adjusted concentrations; see Kapustka et al. 1995) as well as extrapolations across taxonomic groups. Nevertheless, the appropriate methods for the hazard assessment portion of a risk assessment (i.e., phytotoxicity) are fundamentally the same as those for the proposed EU hazard classification framework.

---

[1] These are not germination requirements per se; they are actions required to break dormancy.

Contaminated sites

Contaminated sites invoke variations of hazard assessment concepts. Risk assessment, as referred to above, is married with forensic ecology investigations and remediation options (Fairbrother et al. 1997). If phytoremediation is feasible for clean up of site contaminants, a type of hazard assessment that emphasizes identification of the most tolerant plant species (rather than the most sensitive species) is undertaken. The basic testing procedures are the same as those for hazard classification and risk assessment, except, of course, the selection of test exposure concentrations is directed at the highest levels of tolerance.

Risk assessment of sites known to be, or potentially, contaminated usually involves the use of some form of site-specific risk characterization. This might involve the use of site assessment data and most often includes data for different environmental matrices (e.g., soil, ground water, and possibly surface water). Additional data may be needed to refine or strengthen risk assessments for contaminated sites. These data include biological endpoints (animal and plants) that are specific to relevant exposure pathways.

Objectives

The primary objective was to develop a technically sound approach to generate hazard information about plants for use in hazard classification and ERA. The specific guiding objectives were

- to establish a framework for conducting tests that would be relatively easy to administer, have high accuracy and low variability;
- to capture a range of sensitive endpoints that include measures of survival and growth for short-term tests and survivorship, growth, and yield for longer-term tests;
- to identify technical limitations of information for different uses (e.g., hazard classification and risk assessment); and
- to recommend research efforts needed to advance the utility of data for hazard classification, risk assessment, and site remediation.

Overview of Existing Tests

Phytotoxicity tests have been used to evaluate the level of hazard to plants caused by naturally occurring substances, and both organic and inorganic synthetic substances. Results of these tests have been used for registering pesticides, establishing limits for release of industrial chemicals, and evaluating contaminated sites. Because of the fundamentally different questions being addressed for different regulatory purposes, technical guidelines governing tests have become more focused. Variations in species tested, study design, endpoints measured, statistical analysis, quality assurance-quality control (QA-QC), and reporting requirements occur among the tests that have been standardized for regulatory purposes (OECD 1984; Holst 1986a, 1986b; USFDA 1987a, 1987b; Greene et al. 1988; APHA 1992; ASTM 1994, 2001).

Characterizing adverse conditions at contaminated sites stimulated interest for toxicity data on an expanded suite of species. Though the standardized tests collectively list 31 agronomic species suitable for toxicity testing, the open literature contains reports on phytotoxic responses to chemicals for more than 1500 species from nearly 150 families (Kapustka 1997). American Society for Testing and Materials (ASTM) (1994) produced a standard guide for conducting early seedling growth toxicity tests. An expanded standard guide for conducting terrestrial plant toxicity tests (ASTM 2001) established the parameters for testing a wide range of species, including perennials, woody species, and other nondomesticated plants. Procedures also are described for testing either amendments to soil or site soils that have varying levels of contaminants.

The Organization for Economic Coordination and Development (OECD) currently is revising the "Terrestrial Plants, Growth Test" (OECD test nr 208). The new guidelines will contain greater detail on test conditions, introduction of test substance, and tier testing than does the 1984 version. Guidance on testing also has been produced by the International Organization for Standardization (ISO) (1993, 1995). These tests were designed primarily to investigate effects of organic compounds, especially pesticides, on nontarget plant species. They are limited with respect to their application to contaminated site soils.

Standardized test methods have varying degrees of specificity with respect to test parameters. Generally, the standards have allowed maximum flexibility so study designs can be tailored to specific scenarios to be applicable for regional or site-specific application. There is a significant disadvantage presented by these flexible designs for purposes of generating information for hazard classification. In particular, the acceptable range of pH, organic matter (OM), and clay content in the OECD guidelines would permit testing under markedly different bioavailability conditions. Very different phytotoxic effect levels across the spectrum of pH-OM-clay can be envisioned. The newest ASTM standard guide directs the principal investigator to specify the objectives of the test and to establish data quality objectives to maximize the utility of the tests. Therefore, for purposes of hazard classification, it is recommended that a relatively narrow range of test matrix conditions be used to minimize interlaboratory variation in test results. A control of these parameters affecting bioavailability also should be included in the different regulatory frameworks mentioned above. Most of the existing standard tests cannot be directly applied to the evaluation or assessment of contaminated site soils without procedural modifications.

Methodology for Assessment

Duration of plant tests can span different stages of a plant's life cycle, including
- germination and emergence,
- seedling growth (characterized by a shift during the early days from the use of seed energy reserves to soil nutrients),
- advanced vegetative stages, and
- reproductive stage (flowering, fruit, and seed setting).

Most plant test methods using a soil matrix as the source for exposure are acute tests focused on the early stages of plant development. Only one full life-cycle test, the Brassica life-cycle test (ASTM 1999), has been standardized, although several modifications with other species are under consideration for test method development (R. Petrie, USEPA, personal communication). If seedling emergence, nondestructive shoot height, development measurements, and survival are recorded during the early periods (e.g., weekly observations from planting time) for use as short-term acute metrics, then the test can continue until chronic metrics can be generated and measured (e.g., yield and seed set).

As noted above, the strict distinction between acute and chronic exposure durations is not relevant, as short-term plant tests provide information that is applicable to the plant's whole life cycle. One underlying scientific principal that makes plant testing different from animal testing is the concept of totipotency of plant cells, tissues, and organs (i.e., the capacity of tissues such as cortex, or organs such as leaves, stems, or roots, to regenerate entire plants). Plants routinely lose substantial portions of their mass (e.g., mowing grasses, pruning hedges, grazing, defoliation by insects, etc.) without detectable changes in reproductive capacity. Also, in many plants, the reproductive organs form in response to internal and external signals from buds, without which the plant would continue to develop vegetative structures. In many species, the signals to switch to the reproductive phase are initiated in response to mild or severe environmental stress. Therefore, greater emphasis on vegetative growth parameters in acute exposure periods than to reproductive organ responses may be warranted. However, many annual plants do not have the root reserves to rapidly regenerate above ground tissue, which can result in an adverse effect on yield. For example, early season frost or herbicide injury to seedling corn plants will result, on average, in a 1 bushel yield reduction for each day of delayed growth. This effect on early seedling growth occurred in sensitive corn hybrids from use of the soil-applied herbicide metolachlor under cool wet soil conditions after a safener was added to the formulation. The standardized Brassica life-cycle test may be used to generate both acute and chronic exposure response data for vegetative and reproductive endpoints.

Hazard assessment for regulatory purposes

Data from toxicity tests can be used for identifying the relative hazard of both soluble and sparingly soluble substances such as metals and metal compounds for regulatory programs, such as pesticide registration or hazard classification and labeling. Such tests should be designed to assess the intrinsic toxicity of the substance and should be sensitive, reproducible, and relatively easy to perform. Therefore, the toxicity tests require a high degree of standardization in terms of procedures, conditions, and test species. It is recognized that this degree of standardization is a major advantage within the regulatory framework of hazard classification. Greater flexibility in test design is needed for other regulatory frameworks such as ecological risk assessment, forensic ecology, or phytoremediation studies for contaminated sites.

Standard test guides designed for assessing toxicity of pesticides and industrial chemicals to nontarget plant species, and for use in site investigations, currently are available for plants. These tests include an early seedling growth test (ASTM 1994), seedling emergence and root elongation tests (OECD 1984, 2000; USFDA 1987a, 1987b; USEPA 1996; ASTM 2001) and a life-cycle test (ASTM 2001). These existing standards, especially ASTM E-1963-98, can be used to determine phytotoxicity in acute or chronic exposures for soluble metals and sparingly soluble substances (including metals or organic molecules) for hazard classification.

Soil matrix

One of the most important variables influencing the relative hazard assessment of substances is the test matrix itself. The need to minimize interlaboratory variability of toxicity test results, dictates a narrow set of soil parameters for hazard identification tests. Use of an artificial soil for all hazard determinations provides the best opportunity for ranking, or making relative comparisons, of hazard of all test substances. Moreover, it provides a means of referencing results for tests conducted on complex natural soils including those from contaminated sites.

It is recommended that the test matrix be a standardized solid (e.g., whole-soil) matrix that is easily formulated from "natural" constituents that are readily available to laboratories globally. This will enable laboratories in different regions to access and generate a test matrix that will be relatively similar. Specific guidance on the constituents and formulation procedures will minimize differences among laboratories. The artificial soil originally proposed by the OECD (1984) for use in toxicity tests with earthworms provides the basis for a standardized test matrix for use in acute (e.g., short-term, 14-day) toxicity tests with plants. This led us to prescribe an artificial soil matrix consisting 75% silica sand, 20% kaolinite, and 5% (peat or O-horizon) OM.

Although the disadvantages of using an artificial soil instead of a natural soil are recognized, it is maintained that the advantages outweigh the limitations for hazard identification. Additionally, the proposed master variables (e.g., pH, texture, OM content, and clay mineralogy) in the artificial soil satisfy the physico-chemical properties representing a "worst-case" (i.e., maximal metal availability) soil matrix for the cationic metals (Chapter 2).

There is increasing support in the scientific community for the use of "natural" soils for hazard assessment. However, there are, to date, limited data on the use of such soils and, to our knowledge, no known comparative study assessing the fate and effects of metals in these recommended reference soils relative to that of the artificial soils been published. Until this comparative study is undertaken and the data evaluated and published, it is our contention that the advantages of using a highly standardized test matrix outweighs the advantages associated with the use of reference soils, particularly in light of the uncertainties and inevitable variability accompanying the use of such a matrix.

The artificial soil can be formulated in the laboratory by acquiring the constituents of silica sand, kaolinite clay, and sieved (2 mm) sphagnum peat, determining their water moisture content, and, on the basis of dry weight (dw), mixing the constituents together in a ratio of 75:20:5% until visually uniform in color and texture. Gradual hydration with a total of 2 L of deionized water in 500 mL increments, with continual mixing, will produce a moist batch of soil at about 30% moisture content. The pH of the soil can be determined using a soil slurry method (Anonymous 1992) that specifies a ratio of 1:2 (e.g., 20 g soil wet weight to 40 mL deionized water). If the pH of the soil at this time exceeds the boundaries recommended for the soil matrix (e.g., 5.3 to 6), then calcium carbonate ($CaCO_3$) may be added to increase the pH of the soil. A rate of 10 to 30 g of calcium carbonate per kg peat is recommended. The batch of artificial soil should be "equilibrated" at room temperature for 3 days, at which time the soil pH should be measured again. If necessary, additional $CaCO_3$ may be added and mixed thoroughly with the matrix, and the soil is stored, as is, for an additional 4 days. At the end of the 3 to 7 day "equilibration" period, the pH is measured and, if it is within the acceptable range, the soil can be used in a plant toxicity test. It should be noted that the amount of calcium carbonate required to achieve a pH in the range of 5.5 to 6 depends on the nature (i.e., acidity) of the sphagnum peat or the silica sand. Each time a new bag of either of these constituents is used, it might be necessary to adjust the amount of $CaCO_3$ used in the formulation.

Test soil preparation should begin 5 to 7 days before the organisms are introduced into the soils. The moisture content of the artificial soil should be determined and the appropriate quantity of test substance must be introduced into the appropriate quantity of artificial soil on a dw basis. A batch of artificial soil sufficiently large to satisfy the replicate requirements for a given treatment should be amended with the test substance and the matrix-substance mixture (i.e., the test soil in a given treatment) made as homogenous as possible with minimal mechanical mixing. The test soils for each treatment are divided among the replicate test units (e.g., test containers) and allowed to "equilibrate" for a period between 2 and 7 days, after which test seeds would be planted. A loading rate of 5 or 20 seeds per replicate should be used, depending on the seed size and growth form of the test species.

Species and measurement endpoint selection

The test species recommended for the toxicity tests within the hazard classification framework are alfalfa (*Medicago sativa*), barley (*Hordeum vulgare*), radish (*Raphanus sativus*), and northern wheatgrass (*Agropyron dasystachyum*) or perennial ryegrass (*Lolium perenne*). The rationale for test species selection has been described in detail elsewhere (Stephenson et al. 1997; EC 1998). The species selected meet the following criteria:

- high percentage of seedling emergence,
- short time to seedling emergence,
- good rate of early seedling growth,
- limited variability for measurement endpoints among individuals and replicates,
- discernible concentration-response relationship,
- size of seed facilitates handling,
- root strength enables separation of roots from soil, and
- quality seed stock readily available.

The test species were determined with consideration of the regulatory framework for which the tests were being conducted. Therefore, the plant species comprising a test battery might differ for each of the 3 regulatory frameworks describe herein.

Measurement endpoints for both the range-finding and definitive acute seedling emergence plant tests include seedling emergence, root and shoot length, root and shoot wet mass and dry mass, and total wet mass or dry mass (e.g., root plus shoot wet mass or dry mass). For the acute toxicity range-finding test (e.g., 5 to 8 days post emergence), it may not be necessary to measure all of these endpoints. However, at a minimum, seedling emergence and root and shoot length should be measured. For the definitive, longer-term plant tests (e.g., 14 day post-emergence), it is recommend that all endpoints be evaluated. A procedure to express multiple measurement endpoint responses as a single phytotoxic score may be a useful means of representing hazard assessment data for several species (Kapustka et al. 1995).

Test design and statistical analyses

In addition to biological measurements, physico-chemical measurements of pH and conductivity for soils in each treatment should be made at the beginning and end of a test. Changes in these parameters over the duration of the test can assist in the interpretation of any anomalous results of the tests. The artificial soils also may be characterized for macronutrients (K, Mg, Ca, Na, total N) and micronutrients (e.g., Fe, Zn, Cu), cation exchange capacity (CEC), OM content, total carbon, sodium adsorption ratio, and texture (particle size distribution) to assist in interpretation among tests. It is recommended that exposure concentrations be measured for total metals at the beginning and end of the test to establish concentration-response relationships.

Statistical evaluation for the different measurement endpoints can be conducted using probit regression procedures for quantal metrics (e.g., seedling emergence, survival) and nonlinear regression procedures for continuous metrics (e.g., growth). Probit analysis or nonlinear regression analysis will result in EC50, IC50, or LC50 values. Analysis of variance (ANOVA) procedures can be applied to the acute range-finding data to determine a no-observed-effect concentration (NOEC) and lowest-observed-effect concentration (LOEC), but these statistical endpoints are determined largely by the power of the test and the range of exposure concentrations used and, therefore, are discouraged from use in the definitive test. Moreover, the validity of these statistically derived endpoints has been refuted as they violate scientific strictures (Chapman et al. 1998). ECx (e.g., EC20 or EC50) or ICp (IC20 or IC50), values interpolated from regression procedures are valid scientifically, generally are more precise, and are more useful for regulatory purposes such as hazard classification and ERA.

Objectives of the acute range-finding and definitive tests differ sufficiently to warrant entirely different statistical designs. The objective of the acute range-finding test is to define the range of effective concentrations (i.e., those concentrations that elicit a response ranging from 0 to 100% effect). Using the result of this test, exposure concentrations then can be selected to optimize the test design required to describe a concentration-response relationship in the definitive toxicity tests.

Acute range-finding tests

A balanced ANOVA experimental design is recommended for the acute range-finding test. The range-finding test should consist of a minimum of 6 treatments (i.e., exposure concentrations), including the control treatment, with equal replication across treatments. There should be, at a minimum, 3 replicate test units per treatment, although 4 to 6 replicates would be preferable, as the power of the test would be increased. For each of the test species, 5 or 20 seeds should be planted in each of the test units. The test units should contain a volume of 100 to 200 g test soil wet weight. Water should be added to the test soils as needed to maintain hydration.

Acute definitive tests

An experimental design appropriate for the application of regression procedures is recommended for the definitive acute test. The test should consist of between 10 to 13 treatments, including the control treatment, with unequal replication across treatments. Test unit replicates should be distributed as follows: the control treatment should have 6 replicates, the 4 lowest treatments should have 5 replicates each, the middle 3 treatments should have 4 replicates, and the highest 2 concentrations should have 3 replicates each for a total of 44 test units (i.e., pots). Test soils are prepared as described earlier. For each of the test species, 5 seeds should be planted in each of the test units. The test units should contain a volume of 480 to 500 g test soil wet weight. Water should be added daily (or as required) to maintain the test soils at, or close to, saturation.

Seedling emergence data can be analyzed using quantal assessment methods (probit, moving average, Spearman-Karber procedures) whereas nonlinear regression procedures can be applied to growth metrics (shoot and root length, shoot and root, or total wet mass or dry mass). There basically are 5 models (i.e., linear, logistic, logistic with hormesis, Gompertz [specialized logistic], and exponential) that will describe most concentration-response curves resulting from toxicity tests with metals (EC 1998; Stephenson et al. 2000). Each of the models should be reparameterized to include the ECx or ICp as a parameter, which permits the determination of the concentration of the test substance that would result in any specific percentage reduction in growth. The models also should be reparameterized such that an asymptotic standard error and the 95% confidence interval can be determined for each ECx or ICp. If the data show heteroscedasticity among treatments, they should be weighted with the inverse of the variance of each treatment. The most appropriate model for describing a particular data set can be determined by a visual examination of the scatter plot of the data. However, if more than one equation appears to describe the data well, both equations are applied, and the variance of the residuals then is used to select the most appropriate model. The equation that results in the lowest residual variance is considered to be most appropriate (Ratkowsky 1990). In all cases, graphs of the residuals must be examined to determine whether each model was appropriate for each data set and to determine whether assumptions of normality, homoscedasticity, lack of significant outliers, and lack of correlation between variance of residuals and the estimates of Y are met (Draper and Smith 1981). Residuals usually are not examined statistically because graphical examination is more informative and will almost always illustrate whether assumptions are seriously violated. There are a number of problems associated with the application of nonlinear regression procedures (e.g., selection of the initial starting parameters for the models, colinearity among parameters, heteroscedasticity of variances), and a brief description of how to overcome or minimize these problems is discussed in Stephenson et al. (2000).

Chemical analysis

In Chapter 2, the recommendation is that chemical analysis of the total extractable metal fraction be done at the beginning and end of the test. In some cases, it may be desirable to measure exchangeable or exchangeable plus soluble fractions in soil using accepted reference methods (e.g., Mehlich nr 1 double acid for Zn, Ammonium acetate, NaOH, hot water, Mehlich nr 3 for Zn, Cu, Mn; HCl for Zn; Mehlich-Bowling for Cu; and especially Ammonium bicarbonate-diethylenetriaminepentaacetic acid [DTPA] for Zn, Fe, Mn, Cu, NO_3, K, and P).

Reporting

The report should document all critical features of the study objectives, designs, materials, methods, results, analyses, and interpretations. Reporting and archiving of data files should be done in the spirit of good laboratory practices (GLP) in order to maximize accessibility of data for independent review and reanalysis of data. Specific details for suggested reporting of contents are presented in the Appendix to Chapter 5.

Risk assessment

Soil matrix

Hazard assessment for regional or site-specific risk assessments will have maximum usefulness if the tests are performed with representative soils of the region or site. Test conditions that incorporate bioavailability factors into the study design provide important empirical data that generally will reduce uncertainty that otherwise would be unaccounted for in the exposure portion of the risk equation. If generic hazard information is used, then bioavailability factors should be considered, though they might not be verifiable.

Species and measurement endpoint selection

Toxicity-response relationships among species have been examined for intergenus, interfamily, and interclass taxonomic distance and lab-to-field test conditions (Kapustka and Reporter 1992; Kapustka 1996). There is considerable interest among stakeholders in having toxicity test data on "ecologically relevant" species for purposes of evaluating ecological risk. Any region or large site considered in a risk assessment may have a hundred to more than a thousand terrestrial flowering plant species representing dozens of families. Consequently it is impractical to conduct controlled tests on all groups even at the family level. Typically, a small number of surrogate species are selected to represent dominant life forms (i.e., grasses, legumes, nonlegume forms). Though relatively few toxicity studies have used woody species, ASTM E1963-98 (ASTM 2001) presents guidance on testing woody species. Investigators may propose and justify a suite of species that meet the specific study objectives for a risk assessment. Procedures are presented in the

ASTM guide to assist investigators on the use of native species as well as commercially available rangeland and crop species.

Test design and statistical analysis

The specific objectives of the ERA will govern the type of study design. In addition, existing hazard data or other information in the open literature may lower the amount of new testing required to satisfy the ERA. In most cases, site-specific features that influence exposure parameters would be important targets for specific tests. Other tests may answer questions about comparability of operational units or other identifiable areas within the ERA boundaries.

Chemical analysis

The same analyses as recommend for hazard classification (see section Hazard assessment for regulatory purposes) should be conducted for ERA testing as well. Standard analytical methods and compliance with GLPs (or their equivalent) should be followed whenever possible.

Reporting

Report guidelines suggested in the Appendix to Chapter 5 should be followed.

Site evaluation

This section discusses aspects of standardized soil test methods that are important for the evaluation of contaminated sites. Contaminated site evaluation and toxicity testing or verification of cleanup technologies (including phytoremediation) using plants are discussed. The use of plants for contaminated site soil assessment, before and after cleanup or remediation efforts, is the next step in the evolution of toxicity testing with plants. Standardized methods and plant species for hazard assessment and risk assessment of metals have been proposed by ASTM (1994, 2001) and OECD (2000). The objectives of a contaminated-site, toxicity assessment can be similar, or very different, than those of a hazard classification or a nonsite-specific ERA. Objectives for toxicity testing for contaminated site assessment generally include the following:

- Evaluation of a release of a metal or metal compound at an industrial or mining facility, manufacturing site, waste disposal facility, or any environment where a more diffuse release has occurred such as pesticide spray drift or fallout from refinery stack emissions. Plant toxicity tests proposed for hazard assessment and risk assessment would be applicable to evaluate the level of concern or risk in this case.
- Evaluation of the efficacy of remediation technologies, especially where plants form a large role in the cleanup (i.e., phytoremediation). In this case, the standardized tests for hazard classification and risk assessment require modification to account distinctly for different test objectives and endpoint

metrics. The purpose of remediation efficacy testing is to conduct performance tests on plant types that can tolerate elevated concentrations of the target contaminant (metals or organic compounds) and be used to sequester and remove the contaminant from soil and water or to stabilize the contaminant in-place to reduce risk of off-site migration.

• Evaluation of contaminated site cleanup or post-remediation soil and water goals to supplement or facilitate regulatory objectives. The toxicity test objective would be in support of standardized cleanup goals or site-specific cleanup goals, which are derived using risk assessment methodologies.

Efforts to date have been made to standardize as many of the toxicity test procedures as possible, including soil matrix, plant species for testing, endpoint metrics, and analytical procedures to facilitate comparison across different geographical boundaries. The objective of hazard classification is the ranking of the potential hazard obtained under standard toxicity test conditions with clearly defined endpoints. However, contaminated site assessment and remediation efforts require information obtained through direct testing of contaminated soil under conditions consistent with a reduced set of standard test conditions and methods. In the case of contaminated site toxicity evaluation, some flexibility in test procedures is required to account for site-specific conditions as well as differences in regulatory policies.

Subsequent subsections present a proposed framework for evaluation of plant toxicity or tolerance for sites contaminated with metals or other sparingly soluble compounds.

Soil matrix

Metal contaminated sites often are characterized by mixed contaminants (both organic and inorganic) and rarely have uniform contaminant concentrations in the soils across the site. Often, concentration gradients exist at sites that allow for the collection of soil concentration ranges for use in phytotoxicity testing. Site conditions likely will bracket the range of constituent concentrations required to conduct and assess phytotoxicity or plant tolerance.

Assessing soil phytotoxicity may involve 2 approaches:

1) a screening-level assessment that involves the assessment of phytotoxicity based on the highest concentration of the target soil constituent concentration and

2) a response assessment that includes the assessment of phytotoxicity with multiple concentrations or concentration-effect relationships.

Preparation of soil matrix will depend on the purpose of the toxicity tests. If the objective is to assess the contaminant toxicity for site assessment or risk assessment purposes, toxicity tests can be performed with site soils using the preexisting constituent concentrations or with site soils in a dilution series using reference soils or the artificial soil prescribed in section Plant toxicity, tolerance, or efficacy testing to evaluate a cleanup strategy may include the soil matrix preparation described above (e.g., soil dilution methods) or might involve the addition (e.g., spiking) of the contaminant (as a soluble salt) to the remediated soil to evaluate maximum or specific site concentrations. A good understanding of the site conditions is required if site soils are selected for toxicity tests to ensure that existing or potential soil contaminant concentrations are bracketed. Controls can be used to evaluate plant performance metrics in the absence of the contaminant of concern.

Selection of plant species

As with the soil matrix, selection of plant species for testing will depend on the objectives of the site assessment. Screening-level, site-specific, phytotoxicity testing can be conducted using standardized test species such as those recommended for hazard assessment. However, for evaluation of site risks, native or indigenous species are recommended.

Where possible, native or indigenous plant species, or species that are representative of the region, are recommended for phytotoxicity tests. Site conditions might dictate the selection of a plant species for these tests. For example, certain plant species may be present across the site or there may even be preferential species selection based on target constituent concentration gradients at the site. In this case, it is recommended that species found in an area with high and low soil concentrations be tested. In addition to the native or indigenous species, species recommended for the hazard assessment described above should be used to facilitate comparison to other sites and regions.

Nonstandard species or nonnative species are used as test species for evaluation of phytoremediation or cleanup performance. These species are selected based on prior knowledge of their tolerance to the contaminant of concern or because they have attributes that are particularly suitable for remediation and site cleanup.

Test design and statistical analysis

The objectives of the contaminated site toxicity evaluation will govern the test design, the endpoint measurement, and the type of statistical analysis. For example, the test design proposed for ERAs can be used for simple contaminated site screening assessments. Site-specific toxicity testing for risk assessment or phytoremediation purposes would require a design that is flexible and can accommodate a variety of test species, endpoint metrics, and test durations longer than those for a hazard assessment.

It is recommended that the test be conducted with the use of environmental chambers with standardized conditions (e.g., temperature regime, photoperiod, light intensity, etc.) or greenhouses at the site to simulate more natural climatic conditions. The location of pots and treatments within the chambers or greenhouses should be randomized such that the test units or pots are distributed throughout the chambers or greenhouse and any environmental differences are shared equally among the treatments.

Similar to the test design presented for ERAs, measurement endpoints for both the acute range-finding and definitive seedling emergence plant tests include seedling emergence, root and shoot length, root and shoot wet mass and dry mass, and total wet mass or dry mass (e.g., root plus shoot wet mass or dry mass), senescence, or mortality. Some of these parameters can be monitored throughout the test if the test is conducted for a sufficiently long duration.

Leaf condition may be an important indicator of stress. If deemed appropriate, plants can be monitored for leaf wilt, sclerosis (hardening and yellowing of plant tissue), browning, and leaf drop. Recorded observations will consist of a count of the number of leaves exhibiting any or all of these indicators of poor health.

In addition to biological measurements, physico-chemical measurements of pH and conductivity for soils in each treatment should be made at the beginning and end of a test. Changes in these parameters over the duration of the test can assist in the interpretation of anomalous results. Soils also may be characterized for nutrients (phosphorus, magnesium, calcium, sodium, total nitrogen), CEC, OM content, total carbon, pH, sodium adsorption ratio, and texture (particle size distribution) to assist in comparison among tests. It is recommended that soil concentrations be measured for soluble metals at the beginning and end of the test to establish concentration-response relationships.

Certain meteorological factors such as temperature, relative humidity, and light intensity and duration are expected to have a strong influence on the rates of transpiration exhibited by plants and thus will influence the rates of water uptake as well as soluble and target compound uptake. These parameters should be established for the period of the test to represent the site condition. These can be set to represent site conditions if tests are being conducted in controlled environment chambers. In the event that tests are being conducted in a greenhouse or noncontrolled chambers, monitoring of air temperature and relative humidity should be conducted, preferably with use of continuous recording devices such as data loggers. These data can be used to evaluate plant performance related to prevailing meteorological and climatic conditions.

Treatment application

Generally at contaminated sites, a dilution series (a series of soil concentrations less than those found at the site) is required for testing. Reference soils (uncontaminated soils of similar physical or chemical specifications) can be added to the site-contaminated soil to attain the desired test concentrations. Conversely, target constituents may be added to soils collected from either contaminated or reference areas. These should be added as a soluble salt in solution in the same manner as described above for hazard classification testing. If necessary, the pH of the stock solution may be buffered if groundwater or surfacewater interactions are fundamental to the risk or toxicity evaluation.

Plant tissue

Plant tissue samples (leaves, stems, roots) can be collected from each plant for analysis of target contaminant concentrations. Tissue samples can be collected throughout the test duration or at the termination of the study (or when a plant exhibits signs of serious detrimental health effects). Tissue sampling also can facilitate the characterization of essential plant nutrients in the site soils that may be required as part of the site evaluation or for evaluating the effect of the contaminant on nutrient uptake. The plant and soil contaminant concentration data also can be used to develop bioaccumulation factors for the test species.

Insect and disease infestations

Each plant should be monitored for the presence of diseases or infestations of insect pests. This can be recorded simply as a presence or absence variable on standardized data sheets.

Study duration

The objectives of the contaminated site toxicity testing will govern the duration of the toxicity tests. No set or prescribed test duration can accommodate all types of toxicity or tolerance tests for all species. However, some general guidelines are presented.

Toxicity or tolerance testing for phytoremediation purposes generally requires a longer test duration. For example, if rate kinetics of treatment or removal are required or if seasonal water uptake must be defined, a longer test will provide greater confidence in these estimates. Test duration of one life cycle or even one growing season are desirable for these objectives. If the test cannot be conducted for this length of time, then the test must at least be representative of a significant portion of the growing season or life cycle.

Statistical design and analysis

Data evaluation and statistical procedures described above can be used for most of the contaminated site toxicity tests. A balanced ANOVA experimental design can be used for a wide range of testing applications. These tests can be space, labor, and analytically extensive depending on the number of treatments. A partial factorial design often is a practical way to obtain the most useful data required to meet the study objectives. Standard ANOVA and regression analyses can be used to define treatment effects and determine main treatment relationships. Regression procedures and nonparametric techniques (such as multiple range tests) may be required when evaluating plant growth data. Use of regression techniques require an increase in the number of concentrations tested but can produce reliable results with fewer replicates per concentration, thus resulting in no more effort than required in the ANOVA design.

Chemical analysis

Characterization of soil, plant, and water samples must be conducted to facilitate the evaluation of site-specific study objectives. Screening-level toxicity assessments of contaminated sites will require chemical characterization to be similar (determination of exchangeable or exchangeable plus soluble metal fractions) to that proposed under the hazard assessment test methods (see above). However, if the toxicity or tolerance testing is being conducted to evaluate plant species tolerance to elevated metal concentrations and to determine metal removal by plants, both soluble and total metal analyses should be conducted. Often, regulatory agencies will base phytoremediation performance on data obtained using standard regulatory analytical procedures. In addition, plants that are known to sequester metals from soils in high concentration can alter the chemistry of the soil by releasing organic acids, enzymes, and natural chelators in the root zone to facilitate the removal of sparingly soluble metals. Plant and water characterizations are recommended when the testing is used for remediation purposes to evaluate phytoremediation strategies and performance. These data, along with time-series soil characterization, can be used to calculate mass metal removal efficiencies of plant species.

Reporting

Many of the principles of standardized reporting presented in the hazard assessment discussion and Appendix to Chapter 5 apply to the reporting of test data for contaminated site evaluation. Differences may exist in data interpretation and data summarization, which ultimately depend on the objective of the contaminated site toxicity testing. Reporting requirements of scope, purpose, and test method selection must be described in detail in the report. Presentation of results, such as endpoint metrics, may differ from that of hazard assessment and will depend on the purpose of the contaminated site testing.

Conclusions and Recommendations

For hazard assessment in a classification framework, choice of the soil matrix remains a critical point. Because hazard identification is a process that compares one chemical with another, optimal repeatability of tests is a major criterion. The matrix should be environmentally relevant, but allow high bioavailability of test substance. For these reasons, an artificial soil has been recommended by the plant workgroup where test results are needed for the purpose of hazard classification.

As an alternative to artificial soil, the use of two natural soils with differing characteristics has been suggested (see Chapter 2). However, a comparative database on toxic responses of soil organisms in both artificial soil and natural soils is lacking. It is, therefore, recommended that a comparative study be done, aimed at the evaluation of the repeatability and feasibility of plant toxicity tests on the two standard soils and the artificial soil matrix. This comparative study should be executed by different laboratories in Europe and North America and possibly in other continents where test results are needed.

For risk assessment and contaminated site assessment purposes, natural soil substrates of varying characteristics can be used. They should however, not deviate too much from conditions of the natural environment that is being assessed. Toxicity tests used for risk assessment may be carried out in substrates that allow for varying availability of metals, metalloids, and their respective compounds to biota. However, results obtained should be normalized toward the average conditions of the natural environment for which the risk analysis is made. For such normalization, it is crucial that

- the soil factors that determine metal availability are described in detail and
- the influence of these factors on plant toxicity is assessed in a quantitative way.

Although there is a great deal of information on the influence of soil factors on metal phytotoxicity (e.g., Alloway 1990), this information is scattered and incomplete. Therefore, the plant workgroup recommends the execution of a systematic multivariate study into the quantitative relationships between phytotoxicity and main soil factors defining metal availability in order to establish the algorithm that can be used to normalize lab test data to the conditions of the real environment.

APPENDIX TO CHAPTER 5

Test Report Requirements

General

- Laboratories: The names of the laboratories or institutions performing the test should be included.
- Personnel: Name and title of each investigator, and the name, address, and phone number of the employer must be reported.
- Trial identification number.
- Quality assurance: Indicating control measures or precautions followed to ensure the fidelity of the phytotoxicity determinations; record keeping procedures, and availability of logbooks, skill of the laboratory personnel equipment status of the laboratory or greenhouse, certified of GLPs.

Test material and methods

- Dates: Report the actual dates of the studies including dates of preparation of the substrate mixture, application, equilibration period, planting, and observations.
- Test substance: Identification of the test substance must be provided, including chemical name, molecular structure, and qualitative and quantitative determination of its chemical composition (a.i. percentage, formulation description, adjuvants, and surfactants).
- Relevant properties of the substance tested (physical state, solubility and stability in dilution water, pH).
- For compounds with low water solubility, a solvent can be used to make a stock solution; the stability and the solubility in the stock should be included.
- Untreated controls: Detailed descriptions of containers and plants used as controls for comparisons of toxic effects should be included for each test. Untreated controls should be treated and evaluated in the same manner as the treatments.
- Test organism: Identification of test organisms (genus, species, cultivar, and biotypes) and seed source, seed certifications, and viability.
- Substrate conditions: All the edaphic conditions and characterization including texture, clay content, percent OM, pH, K_d, and K_{ow} values.

- Environmental conditions: For growth chamber and laboratory experimentation, light quality and quantity, air temperature, humidity, photoperiods and thermoperiods, and watering schedules should be reported, method of watering, and amounts of water. For greenhouse experiments, approximate light quantity, high and low daily air temperatures, relative humidity, and photoperiod should be reported.
  - A description of all cultural practice should be included.
  - A complete description of the laboratory, growth chamber, or greenhouse should be included in the report.
- Application: All the test substance application procedures should be included in the report. Chemical concentrations in soils are expressed as dw. The number of selected test concentrations, number of plants per concentration, and the number of replications should be included, along with details of calculations for determining concentration of stock solution and serial dilutions. Methods used for addition of test substances to the soil should be described. The carrier of the substance (if any) should be specified.

Results

- Report percent of emergence, plant height, shoot and root, dw and fresh weight, root length, dead plants, or other growth parameters that may have been measured to ascertain toxic effects of the substance upon the plants with dates of observations.
- Phytotoxicity rating, including a description of the rating system.
- Statistical analysis of the results including a effective concentration effect value (EC).
- Tabulation of results indicating percentage effect level for each species as compared to untreated controls.
- Detailed statistical procedures and assumptions.
- Electronic data should be available.

Evaluation

- An evaluation of the toxic effect of the substance tested should be included in the report.
- All the deviations from specified test methods and procedures and how these deviations could affect results of the test should be included.

References

Alloway BJ. 1990. Heavy metals in soils: Their origins, chemical behavior and bioavailability. New York NY, USA: John Wiley and Sons, Inc. 339 p.

Anonymous. 1992. Soil analysis handbook. Reference methods for soil analysis. Athens GA, USA: Georgia University Station, Soil and Plant Analysis Council, Inc. 202 p.

[APHA] American Public Health Association. 1992. Aquatic plants. In: Standard methods for the examination of water and wastewater. 18th ed. Washington DC, USA: APHA. nr 8220. 81 p.

[ASTM] American Society for Testing and Materials. 1994. Standard practice for conducting early seedling growth tests. In: Annual book of standards. Conshohocken PA, USA: ASTM. E1598-94.

[ASTM] American Society for Testing and Materials. 1999. Standard guide for conducting terrestrial plant toxicity tests. In: Annual book of standards. Conshohocken PA, USA: ASTM. E1963-98.

[ASTM] American Society for Testing and Materials. 2001. Standard guide for conducting terrestrial plant toxicity tests. In: Annual book of standards. Conshohocken PA, USA: ASTM. E1963-98.

Chapman PM, Fairbrother A, Brown D. 1998. A critical evaluation of safety (uncertainty) factors for ecological risk assessment. *Environ Toxicol Chem* 17:99-108.

Draper NR, Smith H. 1981. Applied regression analysis. Ontario, Canada: John Wiley and Sons, Inc.

[EC] Environment Canada. 1998. Development of plant toxicity tests for assessment of contaminated soils. Ontario, Canada: EC. Environmental Technology Centre. Report prepared for Methods Development and Application Section. 75 p.

Fairbrother A, Kapustka LA, Williams, BA, Bennett RS. 1997. Effects-initiated assessments are not risk assessments. *Human Ecol Risk Assess* 3:119-124.

Greene JC, Peterson SA, Bartels CL, Miller WE. 1988. Bioassay protocols for assessing acute and chronic toxicity at hazardous sites. Corvallis OR, USA: USEPA. 123 p.

Holst RW. 1986a. Hazard evaluation division standard procedure non-target plants: Seed germination/seedling emergence—Tiers 1 and 2. Washington DC, USA: USEPA, Office of Pesticides and Toxic Substances. EPA-5430-9-86-132.

Holst RW. 1986b. Hazard evaluation division standard procedure non-target plants: Vegetation vigor—Tiers 1 and 2. Washington DC, USA: USEPA, Office of Pesticides and Toxic Substances. EPA-5430-9-86-133.

[ISO] International Organization for Standardization. 1993. Soil quality—determination of the effects of pollutants on soil flora—Part 1: Method for the measurement of inhibition of root growth. Geneva, Switzerland: ISO. 11269-1.

[ISO] International Organization for Standardization. 1995. Soil quality—determination of the effects of pollutants on soil flora—Part 2: Effects of chemicals on the emergence and growth of higher plants. Geneva, Switzerland: ISO. 11269-2.

Kapustka LA. 1996. Plant ecotoxicology: The design and evaluation of plant performance in risk assessments and forensic ecology. In: La Point TW, Price FT, Little EE, editors. Environmental toxicology and risk assessment. Volume 4. Philadelphia PA, USA: ASTM. STP 1262. p 110-121.

Kapustka LA. 1997. Selection of phytotoxicity tests for use in ecological risk assessments. In: Wang W, Gorsuch J, Hughes JS, editors. Plants for Environmental Studies. Boca Raton FL, USA: Lewis Publishers. p 515-548.

Kapustka LA, Lipton J, Galbraith H, Cacela D, LeJeune K. 1995. Laboratory phytotoxicity studies: Metal and arsenic impacts to soils, vegetation communities, and wildlife habitat in southwest Montana uplands contaminated by smelter emission. *Environ Toxicol Chem* 14:1905-1912.

Kapustka LA, Reporter M. 1992. Terrestrial primary producers. In: Calow P, editor. Handbook of ecotoxicology. London, UK: Blackwell Press. p 278-298.

[OECD] Organization for Economic Cooperation and Development. 1984. OECD guidelines for testing of chemicals: Terrestrial plants, growth test. Paris, France: OECD. Guideline nr 208.

[OECD] Organization for Economic Cooperation and Development. 2000. OECD guidelines for testing of chemicals: Soil microorganisms: Carbon transformation test. Paris, France: OECD. Guideline nr 217.

Ratkowsky DA. 1990. Handbook of nonlinear regression models. New York NY, USA: Marcel Dekker, Inc.

Soil and Plant Analysis Council. 1992. Handbook on reference methods for soil analysis. Athens GA, USA: Soil and Plant Analysis Council, Inc. 300 p.

Stephenson GL, Koper N, Atkinson GF, Solomon KR, Scroggins RP. 2000. Use of nonlinear regression techniques for describing concentration-response relationships of plant species exposed to contaminated site soils. *Environ Toxicol Chem* 19:2968-2981.

Stephenson GL, Solomon KR, Hale B, Greenberg BM, Scroggins RP. 1997. Development of suitable test methods for evaluating the toxicity of contaminated soils to a battery of plant species relevant to soil environments in Canada. In: Dwyer FJ, Doane TR, Hinman ML, editors. Environmental toxicology and risk assessment: Modeling and risk assessment. Volume 6. Philadelphia PA, USA: ASTM. STP 1317. p 474-489T.

[USEPA] U.S. Environmental Protection Agency. 1996. Ecological effects test guidelines, seedling emergence tier II. Washington DC, USA: USEPA. OPPTS 850.4225, 712-C-96-363.

[USFDA] U.S. Food and Drug Administration. 1987a. Environmental assessment technical book on seedling growth. Washington DC, USA: USFDA. Technical Assistance Document 4.07.

[USFDA] U.S. Food and Drug Administration. 1987b. Seed germination and root elongation. Washington DC, USA: USFDA. Center for Food Safety and Applied Nutrition, Center for Veterinary Medicine. Technical Assistance Document 4.06.

CHAPTER 6

Summary and Conclusions

Anne Fairbrother, Guy Ethier, Mikael Pell, José V. Tarazona, Hugo Waeterschoot

The convening of this workshop was very timely, as potential hazards of metals and other substances to the soil ecosystem are coming under increasing scrutiny. During the past 5 years, Canada and many European countries have developed soil criteria for assessment and remediation of contaminated lands, and Australia and the U.S. Environmental Protection Agency (USEPA) are working towards this goal. These criteria depend upon information about hazards to plants, soil invertebrates, and soil microorganisms. Additionally, the European Union (EU) is beginning the process of developing a hazard classification system for materials in commerce relative to their potential to cause adverse effects in the terrestrial ecosystem. Standard hazard identification protocols are needed for this effort if the goal of comparative ranking is to be achieved.

The workshop participants recognized the need to standardize methods for hazard identification and testing across all substances, but they clearly identified specific properties of metals that require different approaches than those used for testing organic compounds. Mixing and equilibration of metals in soils during test setup is a major point of difference. The concept of a "transformation protocol" with short-term and long-term tests run before and after the transformation period was cautiously endorsed by the workshop participants, with the caveat that additional work is needed to define the length of the transformation period for various substances, appropriate leaching and storage conditions, etc. The objective of the transformation protocol is to simulate weathering of metals in the environment to determine whether the bioavailability of these naturally persistent compounds might change over time. The use of the transformation protocol would be equivalent to evaluating environmental persistence of organic substances for the purposes of hazard identification.

Soil type also is a large consideration for testing of metals and was discussed at length by the workshop participants. Consensus and closure on this issue were not achieved. It was acknowledged that some widely occurring soils were not represented, in particular soils of a low organic matter (OM) status, such as those found in arid environments, or soils rich in iron and aluminum, such as those in the tropics. The use of a standard, artificial soil matrix was endorsed by the plant workgroup, particularly for the purpose of hazard identification and ranking. The

Test Methods to Determine Hazards of Sparingly Soluble Metal Compounds in Soils. Anne Fairbrother et al., editors.
©2002 Society of Environmental Toxicology and Chemistry (SETAC). ISBN 1-880611-42-2

soil invertebrate workgroup preferred using natural soils that met specifications, such as those developed for EUROSOILs. The use of natural soils also was endorsed by the soil microorganism workgroup, as the proposed tests utilize indigenous organisms rather than the addition of cultures to a soil matrix. The chemistry workgroup also supported using defined natural soils, but pointed out the necessity of different soil parameters for maximizing bioavailability of different types of metals (cationic metals versus anionic metals). Thus, the use of artificial versus natural soil and the number and types of soils required still remain unresolved for hazard identification and ranking. There was general agreement, however, that ecological risk assessment (ERA) requires hazard information developed from natural soils representative of the area under consideration, as our ability to extrapolate toxicity data across soil types is limited.

The plant and soil invertebrate workgroups (Chapters 5 and 4 respectively) proposed standard species and described detailed test guidelines. The soil microorganism workgroup (Chapter 3) recognized that there are no accepted standard methods for determining hazards to microbial communities. Several microbial function tests were recommended, but additional information will be required to standardize the methods and to put the results into an ecological context.

Workshop participants identified many areas where short-term research will be required to formalize test methods. Moreover, although the focus of this workshop was on developing methods to measure effects to organisms from direct soil exposure, it is acknowledged that the overall objective is to evaluate hazard for the terrestrial ecosystem. Therefore, there is a need to identify and suggest further tests for soil ecosystem function, as well as a test for above-ground organisms, such as foliar and aerial invertebrates, and methods for identifying potential hazard to vertebrates from food-chain exposure or direct soil ingestion. Determination of the potential for metals to bioaccumulate in the food chain is much more complex than for synthetic organic compounds, as organisms have evolved various mechanisms to use, exclude, or take up these naturally occurring substances. Bioaccumulation measurements (such as tissue residues) are considered suitable for assessment of bioavailability of metals in soils and transfer of metals in the food chain; however, this falls outside the scope of hazard identification and discussions at this workshop. Workshop participants strongly endorsed the concept that measurement endpoints chosen for all tests should be ecologically relevant for both acute and chronic effects.

It became obvious from the workshop discussions that development of standardized test methods for hazard assessment has reached different stages in soil chemistry, soil microbiology, soil invertebrates, and plants. However, since soil quality is an integrated function of all 4 of these properties (as well as soil physics, which was not discussed at this workshop), the development of soil toxicity tests for each of these disciplines must occur collaboratively. Moreover, since conducting all the proposed tests during a hazard identification program will generate large data sets, effective methods for evaluation of their ecological relevance must be developed. Important information may be missed if each test is evaluated in isolation. This suggests the need for a coordinated research program to develop an integrated strategy for hazard assessment of metals and metal compounds in terrestrial ecosystems.

Acronyms

ADP	Adenosine diphosphate
a.i.	Active ingredient
AMP	Adenosine monophosphate
ANOVA	Analysis of variance
APHA	American Public Health Association
ASTM	American Society for Testing and Materials
ATP	Adenosine triphosphate
CEC	Cation exchange capacity
CV	Coefficient of variation
DDT	Dichlorodiphenyltrichloroethane
DNA	Deoxyribonucleic acid
DTPA	Diethylenetriaminepentaacetic acid
dw	Dry weight
EC50	Effect concentration for 50% of the organisms
ECx	Effect concentration for x% of the organisms
EDTA	Ethylenediamin-N,N,N',N'-tetraacetic acid
ERA	Environmental risk assessment
EU	European Union
FAME	Fatty acid methyl ester
FAO	Food and Agricultural Organization
fw	Fresh weight
GIR	Glucose-induced respiration
GLP	Good laboratory practice

ICp	Inhibition concentration percent
ISO	International Organization for Standardization
IUPAC	International Union of Pure and Applied Chemistry
K_d	Soil partition coefficient
K_{ow}	Octanol-water partition coefficient
LC50	Lethal concentration to 50% of test population
LOEC	Lowest-observed-effect concentration
NOEC	No-observed-effect concentration
OECD	Organization for Economic Cooperation and Development
OM	Organic matter
PLFA	Phospholipid-ester linked fatty acid
PNEC	Predicted no-effect concentration
QA	Quality assurance
QC	Quality control
SETAC	Society of Environmental Toxicology and Chemistry
SIR	Substrate-induced respiration
SLM	Signature lipid markers
SOP	Standard operating procedure
USEPA	U.S. Environmental Protection Agency
WHC	Water holding capacity

Index

A

"Artificial and accelerated aging," 42

Accelerated aging of soil, 42

Actinomycetes, 17, 23. *See also* Microbial toxicity tests

Acute *vs.* chronic exposure, 60, 64

Adaptations
  to alkaline soil, 10
  to metal toxicity, 3. *See also* Sequestration, of metals
  acclimation, 47
  active exclusion, 1, 84
  genetic, 47
  resistance, 20
  and timeframe of dose, 30
  tolerance, 47, 50, 61, 75

Adenosine diphosphate (ADP), 22

Adenosine monophosphate (AMP), 22

Adenosine triphosphate (ATP), 21, 22

Adsorption, 6

Agrochemicals
  fertilizers, 29
  and microbial toxicity tests, 26, 29
  organic compounds in EUROSOILs, 40
  pesticide registration, 65

Agropyron dasystachyum (northern wheatgrass), 67

Alfalfa. *See Medicago sativa*

Alkaline trap, 18, 30

Aluminum, 9

American Society for Testing and Materials (ASTM)
  Brassica life-cycle test, 60, 64, 65
  E-1963-98, 65
  early seedling growth test, 63, 65
  enchytraeid reproductive test, 52
  whole-plant toxicity test, 2

Ammonium, 27, 28

Ammonium bicarbonate, 70

Amylases, 21, 22

Anaerobic environments, 8

Analysis of variance (ANOVA), 43, 68, 75–76

Anionic metals
  bioavailability, 9–10, 11
  removal of excess, 41

Annual plants, 64

Aqueous speciation, 6

Arsenic (V), 40

"Artificial and accelerated aging," 42

Artificial soil
  advantages and disadvantages, 9, 38, 65–66, 83–84
  extrapolation to natural soils, 9, 30
  for plant toxicity tests, 65, 66, 68
  preparation methods, 66

Arylsulfatases, 21, 22

ASTM. *See* American Society for Testing and Materials

Australia, 83

Azur, 24

B

Background levels
  challenges posed by metals, 5–6
  in EUROSOILs, 39
  in test soils, 45, 46, 47

Back titration, 18

Bacteria, 17, 23. *See also* Microbial toxicity tests

Bait laminas, 49, 50

Barley. *See Hordeum vulgare*

Basal respiration, 19, 20, 25, 27

Benomyl, 51

Betanal, 53

Beta vulgaris (Swiss chard), 6

Bioaccumulation
  effects on mammals and birds, 50
  estimation by tissue residues, 84
  invertebrate toxicity tests, 49

Bioavailability. *See also* Chemical speciation; Oxidation state; Sequestration
  cationic *vs.* anionic metals, 9–10, 11
  contaminant pools, 8, 12
  estimation through extraction procedures, 7
  soil factors which affect, 2
  and soil matrix in plant tests, 70

Biodiversity
  assessment, 20
  measurement, 22–25, 29
  three levels of variability, 49

Biologische Bundesanstalt für Land- und Forstwirtschaft (BBA), earthworm toxicity test guidelines, 51

Biolog substrate, 19, 20

Bioluminescence, 24–25

Dose-response function
  bioavailability issues, 5–6
  effects of transformation and leaching, 14
  essential nutrients, challenges of metals, 6
  in test design, 41, 43–44, 72
  units, for hazard evaluation protocol, 11
DTPA (diethylenetriaminepentaacetic acid), 7
Duration of tests
  invertebrate, 44, 51, 52, 53, 54
  microbial, 26, 27
  plant, 61, 75
Dyes, to trace electron transport activity, 22

E

Ecological aspects
  background levels and ecoregions, 39
  diversity of ecosystems and landscapes, 49
  forensic investigations, 62
  function *vs.* community structure, 25, 50
  mesocosm-type experiments, 31, 49
  role of soil invertebrates, 37, 42
  role of soil microbes, 17, 25
  trophic levels, 17, 31. *See also*
    Bioaccumulation
Ecological risk assessment, 61, 84
EC10
  in carbon transformation test, 27
  in range-finding tests, 43
EC50
  in carbon transformation test, 27
  in hazard identification tests, 44
ECx values
  calculation, 44
  in seedling emergence tests, 69
  in soil selection for tests, 9
  and use of Biolog substrate, 20
EDTA (ethylenediamin-N,N,N',N'-tetraacetic
    acid), 7
Effective dose, 5–6
Eisenia andrei (earthworm)
  carbendazim control compound, 47
  description of toxicity test, 51
  OECD protocol, 2, 37
  recommended guidelines, 45
  recommended pH, 46
  test duration, 44
Eisenia fetida (earthworm)
  description of toxicity test, 51
  OECD protocol, 2, 37

Electron transport activity, 22
Emissions, environmental, 3. *See also* Leaching
Enchytraeus albidus (potworm)
  artificial aging of soils with, 42
  carbendazim control compound, 47
  description of reproduction test, 52
  recommended guidelines, 45
Endpoints
  and duration of transformation period, 15
  fungal toxicity tests, lack of, 24
  invertebrate toxicity tests, 44
  measures of microbial function, *vs.*
    community structure, 25
  nitrogen transformation test, 27
  plant toxicity tests, 60, 67, 70–71, 74
Environmental fate, long-term, 15
Enzyme-driven processes, 19, 21–22
Equilibration
  of artificial soil, 66, 67
  recommendations, 26
  of test substance in soil, 8, 11
  and timeframe for evaluation, 12–13, 14
Essential nutrients
  copper, 1, 22, 70
  dose-response curve, challenges, 6
  limitation of value for assessments, 1, 49
  need for physiological ranges, 50
  zinc, 1, 70
Ethylenediamin-N,N,N',N'-tetraacetic acid
    (EDTA), 7
European Union (EU)
  hazard assessment requirements, 60
  hazard classification system, 83
  materials labeling, 2
EUROSOILs, 10, 39–40, 83, 84
Exposure
  acute *vs.* chronic, 60, 64
  dermal, 37
  ingestion, 5, 7, 37, 45
  and plant root activity, 22, 61
  routes of, 3, 42
Extraction
  chemical, 7
  of microbial biomass
    in biodiversity measurement, 23
    DNA techniques, 23–24
    fumigation, 19, 20–21
    sonication, 23
    surfactants, 23
  of nitrogen compounds, 27, 28

SETAC Titles

Avian Effects Assessment: A Framework for Contaminants Studies
Hart, Balluff, Barfknecht, Chapman, Hawkes, Joermann, Leopold, Luttik, editors
2001

Bioavailability of Metals in Terrestrial Ecosystems: Importance of Partitioning for
Bioavailability to Invertebrates, Microbes, and Plants
Allen, editor
2001

Ecological Variability: Separating Natural from Anthropogenic Causes of
Ecosystem Impairment
Baird and Burton, editors
2001

Guidance Document on Regulatory Testing and Risk Assessment Procedure for
Protection Products with Non-Target Arthropods
Candolfi, Barrett, Campbell, Forster, Grady, Huet, Lewis, Schmuck, Vogt, editors
2001

Impacts of Low-Dose, High-Potency Herbicides on Nontarget and Unintended
Plant Species
Ferenc, editor
2001

Risk Management: Ecological Risk-Based Decision-Making
Stahl, Bachman, Barton, Clark, deFur, Ells, Pittinger, Slimak, Wentsel, editors
2001

Development of Methods for Effects-Driven Cumulative Effects Assessment Using
Fish Populations: Moose River Project
Munkittrick, McMaster, Van Der Kraak, Portt, Gibbons, Farwell, Gray, authors
2000

Ecotoxicology of Amphibians and Reptiles
Sparling, Linder, Bishop, editors
2000

Environmental Contaminants and Terrestrial Vertebrates: Effects on Populations,
Communities, and Ecosystems
Albers, Heinz, Ohlendorf, editors
2000

Evaluating and Communicating Subsistence Seafood Safety in a Cross-Cultural
Context:
Lessons Learned from the *Exxon Valdez* Oil Spill
Field, Fall, Nighswander, Peacock, Varanasi, editors
2000

Evaluation of Persistence and Long-Range Transport of Organic Chemicals in the
Environment
*Klecka, Boethling, Franklin, Grady, Graham, Howard, Kannan, Larson, Mackay,
Muir, van de Meent, editors*
2000

Multiple Stressors in Ecological Risk and Impact Assessment: Approaches to Risk
Estimation
Ferenc and Foran, editors
2000

Natural Remediation of Environmental Contaminants: Its Role in Ecological Risk
Assessment and Risk Management
Swindoll, Stahl, Ells, editors
2000

Guidance Document on Higher-tier Aquatic Risk Assessment for Pesticides
Campbell, Arnold, Brock, Grandy, Heger, Heimbach, Maund, Streloke, editors
1999

Ecotoxicology and Risk Assessment for Wetlands
Lewis, Mayer, Powell, Nelson, Klaine, Henry, Dickson, editors
1999

Endocrine Disruption in Invertebrates: Endocrinology, Testing and Assessment
DeFur, Crane, Ingersoll, Tattersfield, editors
1999

Linkage of Effects to Tissue Residues: Development of a Comprehensive Database
for Aquatic Organisms Exposed to Inorganic and Organic Chemicals
Jarvinen and Ankley, editors
1999

Multiple Stressors in Ecological Risk and Impact Assessment
Foran and Ferenc, editors
1999

Reproductive and Developmental Effects of Contaminants in Oviparous
Vertebrates
DiGiulio and Tillitt, editors
1999

Restoration of Lost Human Uses of the Environment
Cecil, editor
1999

Advances in Earthworm Ecotoxicology
Sheppard, Bembridge, Holmstrup, Posthuma, editors
1998

Ecological Risk Assessment: A Meeting of Policy and Science
Peyster and Day, editors
1998

Ecological Risk Assessment Decision-Support System: A Conceptual Design
Reinert, Bartell, Biddinger, editors
1998

Ecotoxicological Risk Assessment of the Chlorinated Organic Chemicals
Carey, Cook, Giesy, Hodson, Muir, Owens, Solomon, editors
1998

Principles and Processes for Evaluating Endocrine Disruption in Wildlife
Kendall, Dickerson, Geisy, Suk, editors
1998

Radiotelemetry Applications for Wildlife Toxicology Field Studies
Brewer and Fagerstone, editors
1998

Sustainable Environmental Management
Barnthouse, Biddinger, Cooper, Fava, Gillett, Holland, Yosie, editors
1998

Uncertainty Analysis in Ecological Risk Assessment
Warren-Hicks and Moore, editors
1998

Atmospheric Deposition of Contaminants to the Great Lakes and Coastal Waters
Baker, editor
1997

Chemical Ranking and Scoring: Guidelines for Relative Assessments of Chemicals
Swanson and Socha, editors
1997

Chemically Induced Alterations in Functional Development and Reproduction of
Fishes
Rolland, Gilbertson, Peterson, editors
1997

Ecological Risk Assessment for Contaminated Sediments
Ingersoll, Dillon, Biddinger, editors
1997

Life-Cycle Impact Assessment: The State-of-the-Art 2nd edition
Barnthouse, Fava, Humphreys, Hunt, Laibson, Moesoen, Owens, Todd, Vigon, Weitz, Young, editors
1997

Public Policy Application of Life-Cycle Assessment
Allen and Consoli, editors
1997

Quantitative Structure-Activity Relationships (QSAR) in Environmental Sciences
VII
Chen and Schuurmann, editors
1997

Reassessment of Metals Criteria for Aquatic Life Protection: Priorities for Research
and Implementation
Bergman and Dorward-King, editors
1997

Whole Effluent Toxicity Testing: An Evaluation of Methods and Prediction of
Receiving System Impacts
Grothe, Dickson, Reed-Judkins, editors
1996

Procedures for Assessing the Environmental Fate and Ecotoxicity of Pesticides
Mark Lynch, editor
1995

The Multi-Media Fate Model: A Vital Tool for Predicting the Fate of Chemicals
Cowan, Mackay, Feijtel, Meent, Di Guardo, Davies, Mackay, editors
1995

Aquatic Dialogue Group: Pesticide Risk Assessment and Mitigation
Baker, Barefoot, Beasley, Burns, Caulkins, Clark, Feulner, Giesy, Graney, Griggs, Jacoby,
Laskowski, Maciorowski, Mihaich, Nelson, Parrish, Siefert, Solomon, van der Schalie,
editors
1994

A Conceptual Framework for Life-Cycle Impact Assessment
Fava, Consoli, Denison, Dickson, Mohin, Vigon, editors
1993

Guidelines for Life-Cycle Assessment: A "Code of Practice"
Consoli, Allen, Boustead, Fava, Franklin, Jensen, Oude, Parrish, Perriman,
Postlethewaite, Quay, Seguin, Vigon, editors
1993

A Technical Framework for Life-Cycle Assessment
Fava, Denison, Jones, Curran, Vigon, Selke, Barnum, editors
1991

Research Priorities in Environmental Risk Assessment
Fava, Adams, Larson, Dickson, Dickson, Bishop
1987